Philosophy of Engineering and Technology

Volume 27

Editor-in-chief
Pieter E. Vermaas, Delft University of Technology, The Netherlands
General and overarching topics, design and analytic approaches

Editors
Christelle Didier, Lille Catholic University, France
Engineering ethics and science and technology studies
Craig Hanks, Texas State University, U.S.A.
Continental approaches, pragmatism, environmental philosophy, biotechnology
Byron Newberry, Baylor University, U.S.A.
Philosophy of engineering, engineering ethics and engineering education
Ibo van de Poel, Delft University of Technology, The Netherlands
Ethics of technology and engineering ethics

Editorial advisory board
Philip Brey, Twente University, the Netherlands
Louis Bucciarelli, Massachusetts Institute of Technology, U.S.A.
Michael Davis, Illinois Institute of Technology, U.S.A.
Paul Durbin, University of Delaware, U.S.A.
Andrew Feenberg, Simon Fraser University, Canada
Luciano Floridi, University of Hertfordshire & University of Oxford, U.K.
Jun Fudano, Kanazawa Institute of Technology, Japan
Sven Ove Hansson, Royal Institute of Technology, Sweden
Vincent F. Hendricks, University of Copenhagen, Denmark & Columbia University, U.S.A.
Don Ihde, Stony Brook University, U.S.A.
Billy V. Koen, University of Texas, U.S.A.
Peter Kroes, Delft University of Technology, the Netherlands
Sylvain Lavelle, ICAM-Polytechnicum, France
Michael Lynch, Cornell University, U.S.A.
Anthonie Meijers, Eindhoven University of Technology, the Netherlands
Sir Duncan Michael, Ove Arup Foundation, U.K.
Carl Mitcham, Colorado School of Mines, U.S.A.
Helen Nissenbaum, New York University, U.S.A.
Alfred Nordmann, Technische Universität Darmstadt, Germany
Joseph Pitt, Virginia Tech, U.S.A.
Daniel Sarewitz, Arizona State University, U.S.A.
Jon A. Schmidt, Burns & McDonnell, U.S.A.
Peter Simons, Trinity College Dublin, Ireland
Jeroen van den Hoven, Delft University of Technology, the Netherlands
John Weckert, Charles Sturt University, Australia

The *Philosophy of Engineering and Technology* book series provides the multifaceted and rapidly growing discipline of philosophy of technology with a central overarching and integrative platform.

Specifically it publishes edited volumes and monographs in:

- the phenomenology, anthropology and socio-politics of technology and engineering
- the emergent fields of the ontology and epistemology of artifacts, design, knowledge bases, and instrumentation
- engineering ethics and the ethics of specific technologies ranging from nuclear technologies to the converging nano-, bio-, information and cognitive technologies
- written from philosophical and practitioners' perspectives and authored by philosophers and practitioners

The series also welcomes proposals that bring these fields together or advance philosophy of engineering and technology in other integrative ways.

Proposals should include:

- A short synopsis of the work or the introduction chapter
- The proposed Table of Contents
- The CV of the lead author(s)
- If available: one sample chapter

We aim to make a first decision within 1 month of submission. In case of a positive first decision the work will be provisionally contracted: the final decision about publication will depend upon the result of the anonymous peer review of the complete manuscript. We aim to have the complete work peer-reviewed within 3 months of submission.

The series discourages the submission of manuscripts that contain reprints of previous published material and/or manuscripts that are below 150 pages / 75,000 words.

For inquiries and submission of proposals authors can contact the editor-in-chief Pieter Vermaas via: p.e.vermaas@tudelft.nl, or contact one of the associate editors.

More information about this series at http://www.springer.com/series/8657

José Luís Garcia
Editor

Pierre Musso and the Network Society

From Saint-Simonianism to the Internet

 Springer

Editor
José Luís Garcia
Instituto de Ciências Sociais
Universidade de Lisboa
Lisbon, Portugal

ISSN 1879-7202   ISSN 1879-7210   (electronic)
Philosophy of Engineering and Technology
ISBN 978-3-319-45536-5   ISBN 978-3-319-45538-9   (eBook)
DOI 10.1007/978-3-319-45538-9

Library of Congress Control Number: 2016963592

© Springer International Publishing Switzerland 2016
This work is subject to copyright. All rights are reserved by the Publisher, whether the whole or part of the material is concerned, specifically the rights of translation, reprinting, reuse of illustrations, recitation, broadcasting, reproduction on microfilms or in any other physical way, and transmission or information storage and retrieval, electronic adaptation, computer software, or by similar or dissimilar methodology now known or hereafter developed.
The use of general descriptive names, registered names, trademarks, service marks, etc. in this publication does not imply, even in the absence of a specific statement, that such names are exempt from the relevant protective laws and regulations and therefore free for general use.
The publisher, the authors and the editors are safe to assume that the advice and information in this book are believed to be true and accurate at the date of publication. Neither the publisher nor the authors or the editors give a warranty, express or implied, with respect to the material contained herein or for any errors or omissions that may have been made.

Printed on acid-free paper

This Springer imprint is published by Springer Nature
The registered company is Springer International Publishing AG
The registered company address is: Gewerbestrasse 11, 6330 Cham, Switzerland

Contents

1 **Introduction: Towards a Critical Philosophy of Networks – Reflections on the Perspective of Pierre Musso** 1
José Luís Garcia

2 **Network Ideology: From Saint-Simonianism to the Internet** 19
Pierre Musso

3 **Network, Utopia and Fetishism** 67
Filipa Subtil and Pedro Xavier Mendonça

4 **Retiology as Ideological Determinism in the Media: A Political Economy Perspective** 83
David Fernández-Quijada

5 **History, Philosophy, and Actuality of the Utopian View of Technology: On Pierre Musso's Critique of Network Ideology** 103
Steven Dorrestijn

6 **From the Critique of the Network Symbolic Form to the Ideology of Innovation: An Appraisal of Pierre Musso's Work on the Current World Situation of Technology** 131
Francisco Rüdiger

7 **Paradise, Panopticon, or Laboratory? A Tale of the Internet in China** 155
Dazhou Wang and Kaixi Wang

**8 The Rise of Pirates: Political Identities and Technological
   Subjectivities in a Network Society** 187
   Rodrigo Saturnino

9 Final Note: Examining the Network Concept 205
   Pierre Musso

About the Authors 211

Chapter 1
Introduction: Towards a Critical Philosophy of Networks – Reflections on the Perspective of Pierre Musso

José Luís Garcia

One of the most original philosophers to have addressed the world of information and computing technology in his writings, the Frenchman Pierre Musso, took a degree in Philosophy at the *École Nationale Supérieure des Postes et Télécommunications*, before obtaining a doctorate in Political Science at the University of Paris 1 Panthéon-Sorbonne, where he would later teach. His thesis, supervised by Lucien Sfez, was on the topic of the symbolic aspects of telecommunications and the ideas of Henri Saint-Simon (1760–1825), born Claude-Henri de Rouvroy, Count of Saint-Simon. This was the beginning of his theoretical writing on the ideology and history of technical networks and the technological-ideological movement formed by Saint-Simon's disciples. From the body of his work we would highlight *Télécommunications et philosophie des réseaux: la postérité paradoxale de Saint-Simon* (Musso 1997), *Saint-Simon et le Saint-Simonisme* (Musso 1999), *Critique des réseaux* (Musso 2003) and his co-editing of the monumental critical edition of Henri Saint-Simon's *Oeuvres Complètes* (Grange et al. 2012).

The technical network is the key topic and concept in Musso's critical approach to information and communication technologies. He has carried out in-depth research on the dissemination and prevalence of networked power, not only in its instrumental aspects, but also and above all in its cultural and political aspects, adopting an approach involving a multidisciplinary dialogue. Musso sees omnipresent technical and industrial networks in electricity supply systems, transport, inter-bank systems, communication and the Internet – the digital medium *par excellence*. It can be said that technical networks, in their digital forms, cover all domains – the cognitive, the communicational, the cultural, the scientific, the technological, the political, economic and financial, the state, security, leisure... They shape our day-to-day lives, as a technological system on which individual and

---

J.L. Garcia
Instituto de Ciências Sociais, Universidade de Lisboa, Lisbon, Portugal
e-mail: jlgarcia@ics.ulisboa.pt

collective life is increasingly dependent, as well as the most primordial feelings of belonging and identity. As representation and symbol, the technical network is particularly well-suited to understanding that wide-ranging digital world, and the linkages between – or, increasingly, the fusion of – market, technology and science. While its operational impact is clearly not to be understated, it is this metaphorical power which renders the network an ideological, cultural force or, in Musso's words, a "retiology", and as such a favoured topic for critical thought.

Musso's research looks to history, the philosophy of technology and the social sciences in his search for the origins of the idea of the network. He argues that the concept did not arise with the Internet, but has its roots in the ideas – or perhaps it would be more accurate to say in the plans – of Henri Saint-Simon and his disciples who became high priests of industrial development, Saint-Amand Bazard (1791–1832), Barthélemy Prosper Enfantin (1796–1864), and Michel Chevalier (1806–1879), among others. It was reformulated by various figures of what might be called the French nineteenth-century socialist tendency. In France, that group of reformers, industrialists, financiers and above all engineers was charged with the design, planning and promotion of large building projects – roads, canals, railways, telegraph lines and similar projects carried out not only in France, but all over the world. Networks, and in particular transport and banking links, were imagined as material and spiritual bindings which would lead to a "universal association" which would finally tie humanity together and awaken somnolent regions from their torpor. Improving communications was a way of enabling all members of the human family to take part in the ability to travel and explore the world which they had been given as their inheritance. An industrialist vision, in which all sorts of factories would produce all the goods required, would lead to such prosperity that there would be an end to war fever and destruction, and peace would reign in the world. The emerging society would have to be based on industry, which was seen as the guarantee of its material existence, and in doctrinal terms as the only source of all riches and prosperity. Engineers and financiers would be the leaders of such a society.

Pierre Musso is professor of Information and Communication Sciences at the University of Rennes 2 and at *Télécom Paris Tech*, where he holds the education and study chair in "Modelling imaginaries, innovation and creativity" (*Modélisations des imaginaires, innovation et création*). Before becoming a university lecturer, he was a researcher at the *Centre National d'Études des Télécommunications* (CNET) and was in charge of a number of study and research departments at *France Télécom*, such as *Créanet* (today known as *Creative Studio*), the *Institut National de L'Audiovisuel* and the *Délégation à l'Aménagement du Territoire et à l'Action Régionale* (DATAR). Currently he continues to do research work, in institutions such as the *Laboratoire Traitement et Communication de l'Information*, a research centre at the *Centre National de Recherche Scientifique*, and at *Télécom Paris Tech*.

This book takes as its organizing theme "Network ideology: from Saint-Simonianism to the Internet", which is the title of the lecture given by Pierre Musso, as keynote speaker, at the International Conference of the Society of

Philosophy and Technology (SPT), held from 4 to 6 July 2013 at the University of Lisbon, chaired by José Luís Garcia. In addition to this introduction it brings together various researchers' contributions on Musso's theories, and the final chapter contains his brief comments on those contributions. This is the first volume of a collection devoted to discussing keynote contributions to the SPT's international events.

The *Ideologization* of the Network

In "Network ideology: from Saint-Simonianism to the Internet" (Chap. 2, this volume), Musso traces the origins of the idea of the technical network to the current of utopian thought in France in the nineteenth century and, in so doing, critiques the rhetoric which celebrates technical networks, and particularly the Internet, as drivers of technical and social changes which will supposedly have unstoppably benevolent consequences for the future of humanity.

Musso's work enables us to acknowledge him as a remarkable exponent of a tendency in Francophone philosophy and social science – still little known, unfortunately, in the Anglo-Saxon world – involving theorists who have produced work of major importance for understanding the dynamics of the so-called information society, the web and the digital world, like Philippe Breton (1992), Armand Mattelart (1994), Lucien Sfez (1995, 2002), Patrice Flichy (2001), Bernard Stiegler (2006), André Vitalis (Mattelart and Vitalis 2014), and Serge Proulx (Proulx et al. 2005). This list makes no claim to be comprehensive. Notwithstanding their very different individual characteristics, this constellation of thinkers argues, in an already vast bibliography, that in the twentieth century the hidden world of communication became the *locus* for a modern process of "ideologization", to adopt a broad interpretation of Reinhart Koselleck's (1972) notion of *Ideologisierbarkeit*. Underlying this approach is the idea that technology does not have to be viewed as a force separate from culture.

It should be recalled that France had an admirable precursor in the study of modern information systems and the media, in the person of Gabriel Tarde, and that it was also in France in the 1960s, as pioneers in Europe, that French researchers at the *Centre de Études de Communications de Masses* (CECMAS) and the journal *Communications* – in particular Georges Friedmann, Edgar Morin, Roland Barthes, Abraham Moles, Christian Metz – addressed the subject-matter of communication, the media and the so-called mass culture, ahead of – and later interacting with – British cultural researchers like Richard Hoggart, E.P. Thompson and Raymond Williams. But what distinguishes Musso, and some of the Francophone researchers mentioned above, is that they incline towards a critique, whether subtle or more vehement, of the technological utopianism which evolved around new information and communication technologies as a projection of the redemptive change which would put an end to the curses of developed societies – technocracy, disenchantment, social fragmentation, and total control. Those technologies were rapidly incorporated in the rhetoric with which the modern West celebrates the advent of each new technology. Despite some differences, it is not wholly inappropriate to mention here some antecedents in this particular line of French thought, such as Guy Debord and Jean Baudrillard, who saw mass media as the drivers of, respectively, a "society of the spectacle" and the world of the "simulacrum". The critique of technological utopianism is a valuable approach precisely because it questions both the acceptance of technological utopia as the prevailing "topia" of our time, and the fact that it occupies the whole of the potential utopian imagination.

In this connection it should be emphasised that the francophone tendency to criticize communication utopias can be seen as part of a broader pattern of denunciation of technological utopias. The critique of technological utopias has been much cultivated by the philosophy of technology and the social sciences in France, but it has its influential Anglo-Saxon intellectual antecedents like Samuel Butler, Jonathan Swift, Aldous Huxley and George Orwell, the latter, as is well known, addressing communication and the link between information technologies and political tyranny in his acclaimed novel *1984*. We should also take into account the tradition of Goethe in *The Sorcerer's Apprentice*, which was imaginatively taken up by the German philosopher of technology Günther Anders (1984) – the sorcerer's apprentice is the human figure in modern technological society.

In the modern world retiology, the origins and development of which Musso describes and interprets, merges with that technological utopianism. Tracing its development in archaeological fashion, the author outlines some of the meanings of network as technical system. From Antiquity to the seventeenth century, as a craft technique, in weaving and threading; with the Industrial Revolution, in the form of an extensive self-regulating mechanism, thanks to steam engines, which made it possible to run trains on the railways; and, since the emergence of the computer, as an "intelligent", self-reproducing technique. While this triad of representations of technical systems may have undergone changes, the association of the concept of network with the human body prevailed. Musso emphasizes that the brain and the nervous system have been compared to networks since classical times, and this explains the intersection of reticular techniques and the image of the body. In this sense, tracing the origin of networks involves not just understanding them as technical elements, but also examining the organicist metaphors inherent in them. This is of crucial importance in understanding how deceptive analogies between human beings and machines can be, and how they lead many into the ontological error of making no distinction between them.

Musso claims that networks became formalized and subject to mathematical interpretation in the nineteenth century, when they acquired their modern meaning, being used in the context of hydrography, geology and the building of land fortifications, and becoming widely known in the aftermath of the Saint-Simonians' industrial "Manifesto", authored by Michel Chevalier. In the ideal of this technical-ideological movement, the network would ensure universal association and communication between humans and, ultimately, their communion. The "fertilization" of nature by communication networks was to be part of that plan, which would extend first to the physical territory and then to society as a whole. This was a project led by the Saint-Simonian engineers, who were the prophets and architects of a new technical, industrial, financial and ultimately social system. As has been suggested earlier, this was a modern ideologization, but one in which the concept of ideology does not involve the Marxist "veil of appearance", but rather reflects the imaginary projection of desired outcomes for humanity.

Musso sees Chevalier's interpretation of the Saint-Simonian utopian ideal based on networks as the most influential of all its interpretations. Arguing for general circulation and international trade using communication networks, this disciple of Saint-Simon outlines, in his *Le Système de la Méditerranée,* a cult of the network, which would abolish human struggles and divisions. Networks would not just provide the outline of a plan, they would also be the foundation for action. Connected with each other, both primary and secondary, material and spiritual, they would create a system which, like the natural networks of the human body, would give life to the whole territory.

For Chevalier, networks were not merely technical resources, but instruments for the propagation of a morality which would achieve the transition from conflict to communion. An ideology was being formed, by analogy with the development of transport and financial networks, holding out the promise of a new society governed by liberty and fraternity. Musso sees in Chevalier's ideas an enchantment with the network (which would transform tensions into connections) and a liberal-technocratic vision which finds echoes in the present day, in various ways. Even today, networks are mixed up in the semantics of equality, liberty and democracy.[1]

Musso notes there are other Saint-Simonian interpretations of networks, before stressing that those interpretations all seem to share the same notion of communication networks, which would become the practices of a cult of universal association involving the whole of society, improving it in the light of the presumed perfection of the network itself.[2] He highlights the work of Pierre-Joseph Proudhon and Piotr Kropotkin on how society permeates the network. For these thinkers, society and the technical network were similar in structure or, in other words, the structure of technical networks followed the patterns of social organization. If the network meant "universal association", its architecture represented a type of social and political organization. The effectiveness of networks, and circulation within them, would depend on the dominant political and economic model in any given society. For Proudhon, representations and images of techniques had an important part to play, although they could be faulty, for which reason he stressed the need for management of the technical network and for an economic policy suited to its

---

[1]In their modern interpretation, Chevalier's ideas were adopted by other Saint-Simonians who made plans and projects based on technical networks. Combined with each other, these different typologies formed a social system which was to be regulated by means of mixed policy solutions linking the role of the State with private and public enterprise, uniting industrialism and liberalism.

[2]Saint-Amand Bazard argued that networks were the engines of a policy of social change; Prosper Enfantin advocated a religion of universal communication guided by the power of networks; Chevalier argued for a liberal and technocratic political economy for communication networks.

structure. Proudhon believed in the promise of the new inventions, particularly railways, and felt that the symbolic aspects of the network would disappear as technical networks were gradually disseminated democratically and in decentralizing fashion. All these visions are imbued with aspirations to redemption through the potential of technological progress and industry.

It is worth mentioning in this context that Saint-Simon's ambition, since 1817, had been to rebuild the whole system of moral ideas. He wanted to reconstruct religious morality and replace it with a new "industrial morality", with the aim of achieving and realizing the greatest possible human happiness. He sought to complement the principle of utilitarian interest with the principle of sympathetic feeling. Moreover, he made no distinction between moral and purely economic ends and, in the well-organized society which he advocated, private interests would be spontaneously aligned with the common interest. For Saint-Simon, it was necessary to do more than just condemn, struggle against or repress selfishness: a social organization had to be created to make use of it in the most beneficial way possible. The industrial system, blended with a certain type of utilitarian doctrine – a new kind of utilitarianism, a "sympathetic utilitarianism", along the lines suggested by Chanial (2001) – was for Saint-Simon precisely the form of organization required.

Technology as Function and as Fiction

Unravelling the ideas and projects promoted by the Saint-Simonian movement allows us to see that technology, at least in the modern world, acquired its full force not just in instrumental terms, but cultural as well. In sum, technology is both function and fiction. Technology fulfils functions such as producing, controlling, informing, shortening distances, etc., but this is not all it does. Technology goes beyond the functional because it emerges, or arises, in a broad context of hopeful expectation. It matters little whether those expectations match the field of human experience or not. Musso believes that enthusiasm for networks has persisted down to the present day as a myth of social transformation, but that enthusiasm is empty of historical references and drained of Saint-Simon's aspirations. The promise of social change became a mere technological utopia and social change itself became a fiction proclaimed with the advent of each new reticular invention. It is this process which Musso elects to criticize in particular. The network has become reified as a technology and, as a utopia, has been reduced to a technological utopia. Like a perpetual promise, the network heralds a better future, but that today is nothing more than a perversion of the Promethean legacy of the nineteenth century.

It should not be forgotten that today's Web is the result of the interactions and adaptations of actors, inventors and organizations such as the SAGE program for obtaining instant information and analysis for military security purposes, MIT psychologist Joseph Licklider's argument in favor of the symbiosis of humans and machines, the dissemination of cyberculture by *Wired* magazine and the libertarian-oriented ideals of many Silicon Valley engineers and technologists. The prevalence of the network as retiology in our time reflects a residue of Saint-Simonian idealism among engineers, industrialists and, Musso notes, the scientists who conveyed the notion of social and political organization within the technical network. Retiological discourse uncovers networks in organizations and territories and continues to claim for itself the metaphors provided by the human body (above all the brain and the nervous system). It has become the engine of a technological messianism, of a reticular utopia which reuses old imagery and appropriates philosophical and utopian concepts for itself. Technocratic, economicist (i.e. over-estimating the importance of economic motives in the life of society) and libertarian tendencies are a distortion of the aspirations of the Saint-Simonians.

In order to understand the current ideology of networks and its discourses, Musso outlines the various factors which contributed to the Saint-Simonian ideal before it expanded to become a technological utopia.[3] In the internal linkages between the concepts devised by different thinkers, he summarises the three aspects which explain the powerful impact of networks on the imagination: the temporal nature of a transition to progress, democracy, and modernity; the acceptance of the technical network as natural by means of metaphor, and rationalization as a design which can be interpreted in reticular forms.

Musso then looks at how the discourses of technological utopianism have gained ground and spread in society. In the bio-social vision, technical and biological networks are one. The technical network not only explains how the body works, it actually *is* the body. It is like a nervous system which, given life by reticular innovations, would cover the whole planet. But the network would also be society's nervous system, and as such would give rise to a new society. Musso provides the example of how technological utopianism is present today in the informationalization and digitalization of society, and in the cult of the Internet, which was supposed to undermine the hierarchical structure of pyramid-like organizations and provide the resistance to controls on communication.

The linking of communication networks with the nervous system would encourage comparisons with living organisms. This analogy led to the fusion of logical machine and biological organism. The brain remained, however, the

---

[3]The biotechnological, Galenic-Cartesian association between networks and the body, which "naturalizes" networks and technical solutions; the Leibnizian formulation of the network as logical or rational design, having no *a priori* meaning, but merely establishing connections and parallels; the technical network as the social and technical revolution which disrupts the existing social organization, adapted from Chevalier's theories and revived by the ideologues of modern networked innovations; the Saint-Simonian-inspired network as a technical solution which would provide peace and prosperity; the network as an economic tool for dealing with crisis, defining techno-economic policies and promising a new society; and the network as a structure which embodies in itself a choice of society or policy.

most complex processor of data. As such, the analysis of how it operated could be applied to other systems.[4]

Reticular models and forms make it possible to explain the operation of the body, the brain, cities, the planet. They identify a hidden structure of organization, on which it is possible to act or to intervene. With its role as a mediator between technique and society, the network becomes a heuristic link between different sciences and disciplines and, at the same time, an instrument for transforming society. In the transition from an industrial to an information society, the reticular model replaces the pyramid with a polymorphic structure. Reticular technologies would reorganize disintegrating and atomized societies. It is in this context that retiology has come to mould the discourse of the sociology of organizations and to see the market as the natural order of things: in the fields of management and economics, firms would be networked organizations, based on a living, fluid, biological design like that of an organism.

These concepts and symbols have helped the modern ideology of the web to assert itself. Musso addresses this in his discussion on cyberspace and the network society. Cyberspace, he argues, is seen as the limitless space where brains and computers are connected to each other, globally, by means of the Internet. Territory, the State and politics disappear as all obstacles, barriers and frontiers are dissolved. This vision is close to the idea of the noosphere put forward by another French visionary, the Jesuit priest, palaeontologist and philosopher of nature Pierre Teilhard de Chardin, who prophesied that a collective human intellect would arise as a new stage in human evolution.[5]

In the so-called "network society", the model for the network is the Internet, which Manuel Castells sees as the network of networks, a supposedly liberating technology which drives a social transformation (Castells 2000 [1996]). For Castells technological networks, which are the material foundation of society, enable the emergence of an "informational capitalism" and of a society in which reticular techniques would rebuild and regenerate that which is disorganized and fragmented as a result of a variety of social and political events. Musso sees in this theory a mechanistic view of networks and a surrender to technological determinism. And he criticizes Castells' definition of the technical network, which he sees as impoverishing, as an example of conceptual decline and network fetishism.

In the context of this discussion, Musso might perhaps have highlighted Yaneer Bar-Yam's theory of complex dynamic systems. Making an analogy with the organism, Bar-Yam (1997) looks at human civilization as a complex system, in fact as a "global human superorganism", and develops two arguments along these lines: on the one hand, there has been a transition in the structure of human

---

[4]This theory was supported by mathematicians and psychologists Warren McCulloch and Walter Pitts. We should recall here that it was on the basis of this assumption that the mathematician and scientist Norbert Wiener asserted that organisms and machines share similar regulatory mechanisms as producers of communication.

[5]See, in this connection, Garcia (2015).

organization, associated with recent changes in technology, economy and society, towards greater global interdependence and, therefore, complexity; on the other hand, the increase in complexity involves the erosion of central and hierarchical control. This means that human organizations previously acted in a way which was simpler than the individual, but are now more complex. Nevertheless, for Bar-Yam, the erosion of central control in social and economic systems does not involve the complete disappearance of hierarchical structures in human civilization, because the increase in uncertainty and insecurity may lead to the activation of "functional segregation models".

Musso's reflections on the network might additionally have benefitted from emphasizing the "power of networks" as one of the main distinctive characteristics of the current hegemony of cyber-technology. This is true of so-called "network effects", in which the usefulness of the medium, like the telephone or the fax, and now the Internet, depends on the number of users. The greater the number of persons accessing the Internet or one of its platforms, the more useful it becomes and the greater the number of different purposes to which it may be put. Thus, in those parts of the economy which are tied (or bound) to digitalization, value may also be determined by the number of users of any given product. This is the heart of the so-called "Metcalfe's law": the advantages of firms in the digital industries grow exponentially in line with the quantity of users (Metcalfe 2013). The result of this pattern of development is the tendency for economic concentration in a small number of very large firms – we need only look at the giants of the Internet: Google, Facebook, Amazon, Yahoo... –, a tendency which matches the telecommunications industry's development model in the nineteenth century.

Basically, Musso makes his critique of retiology as a series of discourses and images, both practical and theoretical, which select the network as the origin and instigator of continuous social change and celebrate the achievements of technological utopianism in the circulation and connection of networks. But its utopian aspirations stop with the Internet, thus suspending the Saint-Simonian dreams of political and social transformation. The only utopias with which retiology consoles and seduces society are the technological ones, in the form, it should be added, of contemporary neolatry – the worship of any new technology just because it is new.

Musso is therefore part of the counter-current to idealized visions of the alleged power of information technologies to resolve conflicts of interest and ideas, to achieve the orderly management of things, to ensure abundance, to expand democracy and of course to prepare the ground for the fantasy of a union of consciences. In the versions which echo the old ideals of French socialism, those visions are called upon to put an end to the eternal evils of human collectivities, divided by politics, oppressed by wars and undermined by incomprehension. Only the fear that these visions will produce the opposite effect, namely the anarchy and disorder foreseen by dystopic thinking, casts a shadow over the various successive forms of those ideals.

Technology as Power, and Power Over Technology

Over a century and a half since the emergence of Saint-Simonianism, research shows there have been other critical examinations of the fervent voluntarism of this technological-ideological movement. As argued and highlighted in the work of the Canadian Harold Innis (1998 [1951]) and in the North American James W. Carey's essays (1992 [1989]), the development of transport and communications not only eclipsed time and transcended space, enabling us to reach ever more distant people and places, or giving rise to another type of long-distance relationship of a mental nature, driven by mass communication and more recently by the Internet.[6] It also made it possible for the State to control people politically, and for trade and trade-minded thinking to expand and for propaganda and rumour to flourish.

The exponential development of the means of spatial mobility and symbolic communication, made possible by technological changes in transport, speeded up progressive connections between all human subsets which had hitherto been more or less independent. The eminent sociologist Norbert Elias reminds us that in the eighteenth century the word humanity still connoted an objective which some intellectual figures in the Western world hoped would be realized, one which currently has become in some senses a social fact which marches on, even if not in linear fashion on account of the inherent tensions arising from the problems of inclusion and exclusion which the disconnect between technological development and social processes produces (Elias 1995).

There is no doubt that we are able to cover great distances today in order to travel, exchange goods, texts, data, sounds and images, and this leads many people to believe that we are living the dream of having means of mobility and communication not subject to uncertainty and which will promote harmonious connections between giving and receiving. The condition of endless separation among humans has been finally overcome, or is about to be so. Inebriated by the technological power we have achieved, which derives from the purpose of the taming of nature by humans through some of the main paths taken by Western modernity, we forget the consequences of borders of various kinds, whether easily transcended or not, which we maintain between human beings, with other living things and with nature.

The instrumental power and expectations associated with transport and communications over the last century and a half are identified with the world's representations of the countries mainly responsible for creating and producing high technology, with the USA for the time being at the forefront. That world avidly imports and copies the North American technology and technological culture. The conquest of the West by the stagecoach and the railroad, Ford and the civilization of the automobile, Apollo 11 and going to the Moon, Microsoft, Apple, CNN, CBS, Google and Facebook are celebrated examples of the willpower, capability and projection of the United States.

---

[6]On this subject, see also Subtil (2014).

It should however be borne in mind that in eighteenth and nineteenth century progressive thought, human beings sought to tame nature through technology in order to place it at their service. They were not subordinated to technology, a situation which the increasing control over the natural world made possible by late twentieth-century and early twenty-first-century technologies risks bringing ever closer. The unprecedented scale and terms of that dominion over nature suggest an attempt to uncouple human beings from the natural world and to create new contexts, not only culturally, but for life itself. This process places human beings now as demiurges, now as subordinates of the technological system they have created. The new technologies, intensifying the spatial bias of the modern media, as identified by Innis (1998 [1951]), created a new environment, a new technologically-based social space. It is in this space that the information society thrives and many human activities take place remotely and in networks, whether individually, in groups, or in organizations and involving all forms of power (Garcia 2014).

The importance which Musso, and other authors such as Mattelart (1994), attach to the Saint-Simonian movement should not be underestimated, because the Saint-Simonians outlined a complete worldview, one which has emerged in our own day in the central importance attributed to innovation and technological progress, as well as to industry, science and technology leaders. As mentioned at the beginning of this introduction, this is an important tendency in French thought which is little known and cited, at least in the world of the philosophy of technology and engineering. It is understandable, given that the foundations of our time rest largely on various forms of engineering and the entanglement of market, technology and science, that the ghost of their countryman Saint-Simon, as a prophet and theoretician of networks and industrialism, should haunt the minds of influential contemporary French thinkers. But others as well as the French saw that Saint-Simon combined an extraordinary power of prescience with an ability to modify the angle of view at which the perennial issues of philosophy and the human sciences are formulated.

In this connection it is important not to overlook the legacy of Frank Manuel's (1962) view of Saint-Simon as one of the Paris prophets, together with Condorcet, Turgot and others. Particularly enlightening, however, is the essay which the renowned political philosopher Isaiah Berlin (2002) dedicated to this French thinker and visionary. For the former Oxford academic, Saint-Simon is to be regarded as one of the most significant thinkers, if not the most significant thinker, of our time.

From Saint-Simon's legacy to our own day Berlin (2002) highlights plans to reform society by means of large technological projects – he tenaciously pursued the idea of a canal in the isthmus of Panama, to link the Atlantic and the Pacific – the idea of making nature serve humanity, the application of science to production, the veneration of industrialists, bankers and businessmen, and a conception of the State as an industrial enterprise or a kind of business corporation. For Berlin, Saint-Simon should also be recognized as the father of European historicism, before Hegel and anticipating Marx, who gave serious attention to economic factors in history, and the twentieth-century prophets of globalization, an indefatigable defender of the construction of physical ways of uniting humanity.

As a contribution to the debate which this book seeks to encourage we also need to mention a key issue which Berlin does not address, but which is at the heart of Saint-Simon's problems and has become much more crucial in our time: the complex interaction between politics and science and technology, which is reflected in a duality which has left a clear mark on the whole of the twentieth century and the twenty-first century until now – science and technology as power, and political power over science and technology. The relationship is crucial because science and technology, particularly when associated with industry, help to transform our lives, our ways of thinking, and power in social life; and because the political sphere, recognizing the possibilities of science and technology, seeks to subject it to guidelines defined in the light of its visions and political, economic and military interests, etc. Science and technology are recognized as significant elements of power, which makes it understandable that power should seek to guide or even determine the scientific and technological sphere. This is something which political philosophy and political science have largely neglected, with exceptions like Berlin in relation to Saint-Simon, at a time when politics has increasingly become an exercise in technological management, given its relationship with economic growth and financialization. That is why Berlin perceives Saint-Simon to be the principal modern prophet of the technological society.

The Internet today is the most complex technological system in the world and in historical time. Even though it has only existed for a relatively short time, it has led significant thinkers, scientists and critics to produce a long list of wide-ranging diagnostic and forecasting-type hypotheses, some triumphalist, some pessimistic, others simply critical, in relation to the changes which have been brought about or encouraged by the global extent of its impact. It has led to gradual changes in individual behaviour, both neurological and psychic, in the attention economy, and in language; to changes in day-to-day social life, in how people live and socialise together and apart, in the information economy, in science, in scientific communication, in education, the arts, the judicial system, ways of dealing with the environment; to changes in civic life and urban policies, from flash mobs to alleged e-revolutions, e-government and civil society, or possibly producing notions of a "connected society" and new forms of public policy; to the emergence of a variant, or a new epoch, of the individualism which has become institutionalized in the West, in the form of "networked individualism" (Rainie and Wellman 2014); and to changes in progress which in their overall scope may point to the emergence of new human and anthropological forms, perhaps *homo connexus*, who nonetheless has much in common with the *homo urbanus* prevailing in forthcoming decades (Martins 2005; Martins and Garcia 2013).

Six Distinctive Contributions to the Debate on Musso's Thesis

Let us now summarize the six contributions to the discussion of Musso's theories of networks and retiology. Subtil and Mendonça analyses the technicizing impact of networks on the idea of communication and the influence of the network ideal on the current direction of technology in the service of power and economic advantage. Communication networks, in the Saint-Simonian ideal of the technical network, reduce the distances between classes and peoples, in that they involve people and society. The operation of democracy, itself inherent in and driven by networks, allows this process to take place, as a symbol and vehicle for democracy and equality. Subtil and Mendonça draw on the development of the media to illustrate how the information revolution made possible by networks has become the axis of a new capitalism, and stress the significance of three factors: the consolidation of nations with the introduction of the telegraph; the standardization and industrialization of news procedures and the institutionalization of the press as an engine of power and economic intervention.

With the computer and micro-computing, social life has expanded and knowledge have become disseminated, and the Saint-Simonian ideal of world networks has acquired concrete form in the new era of information. Subtil and Mendonça see this trend in Michel Chevalier. While pointing out that there is no straight line from the Saint-Simonian tradition of technical networks to the current shape of technology as an instrument for seeking power and profit, they demonstrate how the libertarian and liberal ideal present in Saint-Simonianism in the person of Chevalier enables (and advocates) information capitalism.

The irreducible nature of the relationship between technical and political networks is explained by the allusion to a technological utopianism of the Promethean type and by consumption. The utopian disposition is the representational aspect of network dynamics, reflecting the projection of an ideal and the search for its realization. In combining the imaginary and realization, involving individual persons in the attraction of their promises, the network lends itself to idolatry as a symbol of social change and acquires ideological content. The cult of networks and the technological sublime reflect the idea that technology is sufficient unto itself as a political goal.

David Fernández-Quijada's contribution also addresses the influence of the technical network as ideology in today's world. He argues that the modern doctrine of retiology was strengthened and intensified with the advent of the digital era and that, despite the dematerialization of networks, the physical is still very much present in the digital universe.

In the *ethos* associated with the Internet he perceives a materialization of Saint-Simonian utopianism, highlighting peer to peer networks in this context as free sharing communities based on the premise that all members contribute to the operation and maintenance of shared information flows. Fernández-Quijada argues that this model was not, however, created by the new media, nor does it produce egalitarian outcomes, because digital networks themselves produce new hierarchies.

Another example of the way in which the determinism of retiological rhetoric has been reinforced is to be found in the liberalization of the broadcasting industry, with the aim of reusing its frequencies for profit-making mobile telecommunications services. Increasingly, with the changeover to digital television, commercial interests have been favoured over public service interests, and this illustrates the way in which digitalization has accentuated these problems and dynamics.

In connection with the second argument, Fernández-Quijada concurs with the territorial nature of networks, following Musso's suggestion that territory, in addition to its physical aspect, also has a cultural and political existence. He underlines the significance of physical space in the geographical grouping practices (clusters) of industrial firms and of industrial actors associated with production and distribution, and its impact on the geography and characteristics of regions in terms of customers and suppliers, infrastructure and natural and human resources, and low transaction costs as a result of the short distances involved. Rather than arguing that modern retiology can be traced to the earliest stages of the media, Fernández-Quijada demonstrates that retiological determinism is a mirror of capitalist dynamics, just as the threat hanging over public broadcasting companies is a derivative of the neoliberal ethos.

In his article, Steven Dorrestijn outlines the advantages of Musso's contribution, putting together an essay on the utopian, dystopian or ambivalent interpretations of technical mediation, while developing a dual critique of Musso's appropriation of the notions of "network" and "utopia". Dorrestijn sees the breadth of Musso's historical perspective as its principal merit, in that it gives him an analytical advantage when it comes to discussing the issues surrounding technology today. Dorrestijn goes on to explain the origins and meaning of the notion of utopianism and describes the historical development of ideas which link technology and its social worth.

With references to Francis Bacon and Jeremy Bentham, Dorrestijn demonstrates how utopian were Saint-Simon's plans, combining the utopian intentions of technocratic philanthropism with the aim of revolutionising religion. In identifying industry as the desired model for society, the Saint-Simonian project conveys that negative characteristic which Dorrestijn seems to stress as being central to the utopian conception of technology: the lack of critical ethical reflection.

Moving on from the utopian vision, Dorrestijn notes the advent of ethical concerns in relation to technology, before identifying a more recent and ambivalent notion: that technology, deprived of any essence, contains both positive and negative possibilities, so that the way it is implemented becomes significant, and adverse effects can be avoided or corrected. The third part of Dorrestijn's analysis is a critique of this idea. Musso seems to distinguish two sorts of techno-utopianism: one inspired by the Saint-Simon's social semi-utopia, which recognizes the importance of positive technology, and the other which identifies the technical network as the ideal organism, based on Saint-Simonian ideas. Dorrestijn believes that distinction should depend on the relationship between utopianism and social involvement.

If social and political participation depend on utopian inspiration, then perhaps some utopianism may be justified.

In conclusion, he analyses the centrality of the notion of networks in Musso's explanations of techno-utopianism, suggesting that its omnipresence does not necessarily imply acceptance of the techno-utopia. In this connection the work of Bruno Latour is revealing, in that it shows how immersion in the network does not mean abandoning an empirical stance towards concrete social issues. Dorrestijn favours an empirical orientation rather than one subordinated to "mental concepts" and, rather than being critical of Musso's thought, suggests alternatives in the form of a more empirical orientation.

Francisco Rüdiger is concerned with the relationship between retiology and capitalism, but does not limit his critique to this text of Musso's. Looking at other articles and works by Musso, he identifies an integrative and rational view of the idea of technological innovation, which he believes to be problematic and debatable. In Rüdiger's view, Musso argues that through better linkages between technological innovation policies and knowledge of the social imaginary, it would be possible (and desirable) to reconstruct the discourse of the network minimizing its ideological nature, preserving technical rationality and fulfilling consumer desires. Thus Musso not only recognizes the social function of retiology, but also rescues the reticular idea itself and puts forward a rational reworking of the network so that it will contribute to technological innovation. Keeping faith with autonomous reason as the principle which helps us to explain the real world and eventually plan new guidelines for it, Musso acknowledges the mediation of the symbolic and the imaginative on this world. His approach to innovation is an effort to combine older modernist expectations of a rationalization of the culture with the arbitrary nature of social actors' desires today.

According to Rüdiger, Musso's reflections reveal a loss of critical perspective and are covered by a layer of strategic and instrumental interest. He does not address the issue of the will to power which is intrinsic in the capitalist market economy and directs innovation into the channels necessary for greater economic advantage, nor does he investigate the very concept of innovation, and its implementation as a corporate policy at the turn of the twenty-first century.

Along these lines, Rüdiger questions, to a degree, the possibility of shaping the imaginary in an instrumental way, because it is actually the will to power which explains the search for innovations and their eventual adoption. Musso's abandonment of the critique of retiology shows how we are able to criticise Saint-Simonism without escaping the web of the will to power, which is evident in his aim of preparing the ground for the advent of industrial forms capable of shaping the social image of technology for profit-making purposes.

Rüdiger sees Musso as a new philosopher of innovation who, having once criticized retiology, succumbs to a similar ideology, granting academic legitimacy to those who see technological innovation as the key to social change and a replacement for political action. The parallel with Saint-Simonism is clear. The image of innovation is for Musso what the network was for the Saint-Simonians.

Dazhou Wang and Kaixi Wang, and Rodrigo Saturnino enrich the discussion on networks by comparing the ideology of the technical network and Saint-Simonian thought with recent or current political contexts and events. Reviewing how the Internet has penetrated and developed in China, Wang and Wang focus their analysis on how society and the Internet have developed side-by-side. These authors present the Internet as a laboratory, in which social, corporate and governmental actors operate in various ways and engage in power dynamics leading to social change. Wang and Wang highlight the presence of *e-influences*, dissident voices which have given rise to the diversity necessary for social change, allowing for public opinion to be formed among those surfing the Internet.

The right of association, which had been restricted in China, was achieved with the Internet, thanks in large part to Weibo, a microblogging service which encouraged political participation, and led the central government to outline strategies for dealing with the Internet, such as the dissemination of explanations of matters of public interest. This rendered politics more transparent by helping to explain those matters to the general public.

The government, however, fearing threats to social stability, set up the Great Firewall, a system of control and surveillance which blocks certain websites and filters keywords which web surfers key into search engines. Business entities, required to set up security systems to prevent the illegal transmission of information, carry out their own censorship so that web surfers, being aware of the censorship they are likely to suffer online, practice a kind of self-censorship.

In the light of the dynamics resulting from the spread of the web, these authors go along with Latour, who sees the Internet as a laboratory in which social actors, far from having defined and unchanging properties, experiment with its possibilities. With the emergence of personal media, civic participation has triggered mass mobilization in support of specific causes. This, combined with the increase in governmental transparency, contributes to greater freedom and to strengthening the rights to information and publication.

Wang and Wang suggest that there is not one "Internet", but several "internets". This goes against the idea of "universal association" and the conception of the Internet as an automatic conduit for democracy. The authors stress the dynamics of the interaction between the state, society and the Internet, in a process in which all those involved tested the suitability of different behaviours. Their main argument seems to be that social change is not a feature of networks and is not inherent in their architecture, but depends rather on political choices: they lacks, in effect, a political framework which might encourage further exploration of their nature as a laboratory.

Rodrigo Saturnino analyses the emergence and ideology of the Pirate Party of Sweden, whose trajectory reflects the ambiguous nature of the Internet. The main aim of this party, which arose in response to technical and legal attacks on the free sharing of information, was to question the legitimacy of the private sector's drive to monopolize information and thereby restrict civic autonomy.

Its success to date lies in the adoption of a holistic strategy which caters to the needs of different cultural contexts, in line with the basic principles which rely on the technological imaginary. Like Musso with the disciples of Saint-Simon, Saturnino identifies in the Pirate Party's trajectory a utopian inclination based on the reticular imaginary. But the pirates know that to achieve their reticular democratic imaginary they need to institutionalize the struggle and the resistance, by reiterating the libertarian and techno-utopian nature of the network as a democratic instrument. For the pirates, the rhyzomatic nature of the network, which makes control and surveillance possible and provides the means for a new capitalism, also provides the guidelines for a new path of resistance.

Saturnino's contribution is also notable for the way he identifies which forms of network allow for polysemantic interpretations and shows how the uses of the Internet depend above all on their technical structure, even if they have their origins in the reticular ideology and imaginary. The fact that information circulating on the Internet has successfully been placed in the service of the market is a prime example of the polysemantics of technical networks and more specifically of the Internet. A second example of this ambiguity is to be found in the paradoxical relationship between the privatization of information and privacy, inasmuch as the logic of the privatization of knowledge has taken place alongside the adoption of policies which disregard users' privacy. The third example lies in the objectification of contradictory advantages: while on the one hand it encourages the circulation of information, on the other hand it lends itself to control and the institutionalizing of surveillance.

In this sense, because competition for power and the colliding interests of markets and citizens attach to the idea of information and because, in the Pirate Party's philosophy, information is a common good, not only "are we all connected", but "we are all pirates" also.

Pierre Musso himself offers a summary at the end of the book, in which he engages with each of the commentaries, examining the different meanings and symbolisms of networks, and looks at the metaphorical aspects and the models of rationality in which retiology prevails in a variety of spheres today.

Acknowledgement This chapter was supported by the Portuguese national funding agency for science, research and technology (FCT), as part of the UID/SIC/50013/2013 Project.

References

Anders, G. 1984. *Die Antiquiertheit des Menschen: Über die zerstörung des lebens in zeitalter der dritten industriellen revolution*. Munich: Beck.
Bar-Yam, Y. 1997. *Dynamics of complex systems*. Reading: Perseus Books.
Berlin, I. 2002. Saint-Simon. In *Isaiah Berlin. Freedom and its betrayal: Six enemies of human liberty*, ed. H. Hardy, 105–130. Princeton: Princeton University Press.
Breton, P. 1992. *L'utopie de la communication*. Paris: La Découverte.
Carey, J.W. 1992 [1989]. *Communication as culture: Essays on media and society*. New York/London: Routledge.

Castells, M. 2000 [1996] *The information age: Economy, society and culture*, vol. I. *The rise of the network society*. Cambridge, MA and Oxford: Blackwell.
Chanial, Ph. 2001. Les socialismes français. In *Histoire Raisonnée de la Philosophie Morale et Politique*, ed. A. Caillé, Ch. Lazzeri, and M. Senellart, 524–534. Paris: La Découverte.
Elias, N. 1995. Technization and civilization. *Theory, Culture & Society* 12(3): 7–42.
Flichy, P. 2001. *L'imaginaire d'internet*. Paris: La Découverte.
Garcia, J.L. 2014. Une critique de l'économie des communications à l'aune des medias numériques. In *La contribution en ligne: Pratiques participatives à l'ère du capitalisme informationnel*, ed. S. Proulx, J.L. Garcia, and L. Heaton, 49–61. Québec: Presses de l'Université du Québec.
———. 2015. Theodicy. In *Ethics, Science, Technology, and Engineering: A global resource*, 2nd ed. J.B. Holbrook (Editor in Chief), and C. Mitcham (Associate Editor), vol 4: 357–360. Farmington Hills: Gale, Cengage Learning.
Grange, J., P. Musso, P. Régnier, and F. Yonnet (ed). 2012. *Henri Saint-Simon: Oeuvres Complètes*, vol 4. Paris: PUF.
Innis, H.A. 1998 [1951]. *The bias of communication*. Toronto: University of Toronto Press.
Koselleck, R. 1972. Einleitung. In *Geschichtliche Grundbegriffe. Historisches Lexikon zur politisch-sozialen Sprache in Deutschland*. Bd. 1, ed. R. Koselleck, W. Conze, and O. Brunner, XIII–XXVII. Stuttgart: Klett-Cotta.
Manuel, F.E. 1962. *The prophets of Paris*. Cambridge, MA: Harvard University Press.
Martins, H. 2005. The metaphysics of information: The power and the glory of machinehood. *Res-Publica: Revista Lusófona de Ciência Política e Relações Internacionais* 1: 165–192.
Martins, H., and J.L. Garcia. 2013. Web. In *Portugal social de A a Z: Temas em aberto*, ed. J.L. Cardoso, P. Magalhães, and J.M. Pais, 285–293. Paço de Arcos: Impresa Publishing/Expresso.
Mattelart, A. 1994. *L' invention de la communication*. Paris: La Découverte.
Mattelart, A., and A. Vitalis. 2014. *Le profilage des populations*. Paris: La Découverte.
Metcalfe, B. 2013. Metcalfe's Law after 40 years of ethernet. *Computer* 46(12, Dec): 26–31.
Musso, P. 1997. *Télécommunications et philosophie des réseaux: La postérité paradoxale de Saint-Simon*. Paris: PUF.
———. 1999. *Saint-Simon et le Saint-Simonisme*. Paris: PUF.
———. 2003. *Critique des réseaux*. Paris: PUF.
Proulx, S., F. Massit-Folléa, and B. Conein. 2005. *Internet, une utopie limitée: nouvelles régulations, nouvelles solidarités*. Québec: Presse de l'Université Laval.
Rainie, L., and B. Wellman. 2014. *Networked: The new social operating system*. Cambridge, MA: The MIT Press.
Sfez, L. 1995. *La santé parfait: Critique d'une nouvelle utopie*. Paris: Seuil.
———. 2002. *Technique et idéologie: un enjeu de pouvoir*. Paris: Seuil.
Stiegler, B. 2006. *La télécratie contre la démocratie: Lettre ouverte aux représentants politiques*. Paris: Flammarion.
Subtil, F. 2014. Du télégraphe à Internet: enjeux politique liés aux technologies de l'information. In *La contribution en ligne: Pratiques participatives à l'ère du capitalisme informationnel*, ed. S. Proulx, J.L. Garcia, and L. Heaton, 115–125. Québec: Presses de l'Université du Québec.

Chapter 2
Network Ideology: From Saint-Simonianism to the Internet

Pierre Musso

For the last two centuries, each "industrial revolution" in the West has been accompanied by and relied upon the formation of a large territorial technical network: the railways, with the first "industrial revolution" (1780–1830), the electrical network, with the second "industrial revolution" (1880–1930), and finally the Internet network, spawned by the convergence of telecommunication and information technology, with the third "industrial revolution" (since 1960). These major industrial complexes have been defined as "technical macro-systems", for they combine technical networks with power structures (see Gras 1997).

The third industrial revolution, that of information technology and its encounter with telecommunications, has resulted in the generalized computerization of society and the economy, along with the development of the Internet, social networks and information systems, and virtual and digital simulation techniques. The contemporary "technical macro-system" is thus comprised of interconnected information, command and communication networks, interlinked with the transport and energy networks. Many myths, fictions, images and imaginaries[1] have always surrounded the development of major technical networks, with the purpose of socializing them.

A new divinity is tending to prevail today, a technician divinity, and the Internet is but one of its luminous apparitions: "the Network". The figure of the network is becoming ubiquitous. Everything is a network, or even a "network of networks". The organization of daily life becomes a constant use of networks, a quest for

---

All translations are mine, unless otherwise indicated.

[1] We have translated the French "imaginaire" as "imaginary", although the notion is more complex in French. The reader is referred to Gaston Bachelard's philosophical definition.

P. Musso (✉)
Rennes 2 University, Rennes, France

Télécom Paris Tech, Paris, France
e-mail: pierre.musso@telecom-paristech.fr

access or connection to electrical or electronic networks, communication and information networks, urban networks, transport networks, etc., and is fitted into their dense webs covering the entire planet. Commenting on the Network's omnipresence and omnipotence, whether to emphasize its benefits or its threats, has become somewhat trite. Cities become a "Networkopolis" or a "Smart City" resembling a large urban information system, while the Earth turns into a "relational planet". Manuel Castells sees a "network society" emerging and "social networks" are said to define human relations (Bressand and Distler 1986, 1995). The Network even provides interconnected subjects who are "switched-on" with an identity (through Facebook or Twitter). Manuel Castells explains that "our societies are increasingly structured around a bipolar opposition between the Net and the Self" (Castells 2010 [1996]: 3), while philosopher Pierre Legendre notes that "our societies are driven to networked feudalization" (Legendre 2001: 221). Hence, the Network gives meaning and direction. Its effectiveness is enhanced by its mythological foundations, which signal the future and social transition. Social change is now thought to be constantly experienced through connection, being "switched on", digital interaction and immersion in virtual flows and worlds. The technical network thus becomes the end and the means to think and perform social transformations and even present-day revolutions. Be it literary fiction, futurology or the decryption of the network society, the network imaginary is incessantly announcing the "revolution" of (and through) networks. Hence, the digital, Internet, robotic, industrial, and other "revolutions" that are "changing the world" thanks to Apple, Facebook, or Google.

The Network defines not only the new rules of the economy, but also those of power (see Rifkin 2000). At the same time, this constant cult of the Network which is re-enchanting daily life, particularly through the virtues of the Internet, enables us to reinterpret the contemporary world. For the Network has also become a process of reasoning enabling us to think about the world. The unbridled imaginary produced by the network is a product of its embeddedness in technologies; it provides a "techno-imaginary", or even a "techno-messianism", to use anthropologist Georges Balandier's term (2001: 20), and a mode of understanding of the world made all the more powerful by the omnipresence of techniques. The network is at one with techniques, as its entire history attests.

The Tree and the Network

The network is a dual figure. Like the State, with which it is often contrasted, it has the two faces of Janus: the one technical and the other technological, if we agree to consider technology (*tekhné* + *logos*) as a representation and a narrative of technique. The technique-network allows the neo-industrial world to function "efficiently" and the technology-network provides an account thereof. The network is an artefact to amplify action and accelerate movement; it inspires dreams and allows analysis: extraordinary virtues, like those of the tree until the Age of

Enlightenment. Provided by nature, the tree gave a point of reference, it signaled a hierarchical and genealogical order, as well as that of knowledge in the Encyclopedia: from the buried roots to the branches stretching up towards the sky, passing by way of the trunk, the tree distributed filiations and knowledge. The One (the trunk), stemming from the multiple (the roots), again begot the multiple (the ramifications). Through its verticality, the tree ensures the linear transition from the earth to the sky, from the experienced present to the promised beyond. The symbolism of the tree was in a sense "uprooted" during the Enlightenment, with the great overhaul seeking to "disenchant the world". And the re-enchantment was swiftly achieved thanks to the techniques of the industrial world, with its first artificial networks: the "wonderful" railways, the telegraph, the "electricity fairy". The Network has therefore replaced the Tree. The latter's linearity and natural verticality has been opposed to the former's multirationality and apparent horizontality. In Saint-Simonian thought it conjures the equality of the brothers against the hierarchy of the Father. This is one of the factors underpinning its efficiency and power.

The reticular techniques which constitute the infrastructure of hyper-industrialized societies are proliferating and, according to Manuel Castells, seem to outline the structure of a networked "informational capitalism". Simultaneously, the figure of the network is omnipresent in all disciplines, from biology to mathematics, from sociology to political or organizational science, and even claims to define the modalities of thought processes through the cognitive sciences and connectionism. The network, a multidimensional object and fetish word, has become a *doxa* for contemporary thought.

All that remains today are the images and ideologies of the network, but these are the decayed remnants of a social utopia and conceptual thought developed in the early nineteenth century by philosopher and sociologist Henri Saint-Simon (1760–1825), who conceptualized industrial society. We are left with a "technology of the mind" and "a symbolic image" that re-interpret an ancient imaginary of the network with every technical change. This is what we call a *retiology*, a neologism created by contracting *retis* (network in Latin) and *logos*, that is, a set of representations, discourses and images supported by technique-networks.

Archaeology of the Network

The genealogy of the network highlights three major visions of the reticulated in the West, which relate to three technical configurations of the network, emphasizing the indissoluble link between the technique-network and its social representations.

The first and very ancient representation, found in mythology, particularly Greek mythology, relates to thread, fabric and weaving: it is a biometaphysical vision of the network-net symbolizing continuity, the thread of life, time and Destiny.

The second emerged at the end of the eighteenth century, with the formation of a new *episteme* formalizing the network and rationalizing it into a logic which, with the "industrial revolutions", brought in new territorial technical networks, such as railways, the telegraph and then electricity. Saint-Simonianism systematized this second configuration into a biopolitical vision of the reticular in which two political paradigms of the network (centralized/decentralized) are contrasted, and which is driven by a social utopia.

Finally, in the twentieth century, with the computer and information and communication techniques, a third configuration elaborated by John von Neumann and Norbert Wiener emerged, that of the self-regulated techniques symbolic of the brain and of "collective knowledge", all embedded in a "biotechnological" vision of the reticulated. The communication network is thought of as a nervous system or a brain; since Galen in Antiquity (129–200 CE), these organs had been defined with reference to technical networks. Galen saw the brain as a *rete mirabili*, comparable to fishermen's nets. These images and representations between body and technique work both ways. That is why reticular techniques have historically been intertwined with the metaphor of the body: for a long time, from Antiquity to the Enlightenment, the network was "on" and "around" the body; it enveloped the body. At the end of the eighteenth century, the network was identified with the body, and then externalized as an artefact enveloping all of nature, particularly territory. Finally, since the nineteenth century, the body has been entangled in the artificially created technical transport and information networks which constitute its new social environment, maybe even a new society. The network has now cast its nets around society as a whole, as though it has successively enveloped the body, nature, and then society. Memories of these captures have been deposited in strata within the same "network" object, making it possible to circulate from one referent to the other. From Antiquity to the seventeenth century, an imaginary of the network as an inter-world between the weaving technique and the organism developed. At the end of the Enlightenment, this imaginary gave way to a triple rationalization: that of the Promethean productions of the engineers constructing artificial networks; that of the formalization-mathematization inaugurated by Leibniz and Euler, and finally that of the construction of a symbolism of social change meant to materialize through reticular techniques. In its meshing, the textile technique delivered a "graphic reason" to interpret the human body from Antiquity to the Enlightenment. Network and body then faded into a single rationality, a little before the modern technical network found its rationality within the body, from the Industrial Revolution. In other words, for a time the imaginary of the network gave way to the concept in Saint-Simon's philosophy, before deteriorating into an invasive vulgate of the network, an ideology and technology of the mind.

An Inter-world Between Technique and Body

There are two dimensions to the Network, one technical and the other techno-imaginary. A network is first a technique that evolved over the course of history, taking on three main forms: "technical systems" as understood by Bertrand Gille

(1978), that is, a craft weaving technique from which *réseau*, the French word for "network", derives (*retis* in Latin); the major artificial territorial networks that emerged from the industrial revolution; and finally, the information and communication networks that emerged from the information technology revolution.

The network is steeped in an imaginary that is always associated with a technical system. From Antiquity to the seventeenth century, it referred to threads and weaving, to nets or wickerwork, in other words, a crafted form of the reticular. With the industrial revolution, the network became a large self-regulated mechanism thanks to the steam engine that made railways possible, and the technical network was embedded in the territory. Since the mid-twentieth century, with the invention of the computer and John von Neumann's "automata networks", the network has appeared as a self-reproducible technique, even qualified as "intelligent".

Though trilogies structuring history should be considered with caution, it is worth noting that the history of network techniques matches the chronology of the three-phase industrial civilization put forward by Lewis Mumford (1934) in *Technics and Civilization*: until the eighteenth century, the "eotechnics" phase, when network-weaving prevailed; the "paleotechnics" phase of the eighteenth and nineteenth centuries, linked to the industrial revolution, when the major artificial territorial networks built by engineers appeared (transport, energy and communication); and, finally, the "neotechnics" phase that characterizes modern industrial civilization, in which information technology and telecommunication networks have emerged.

Irrespective of the variations in the technique-network concept that characterizes "technical systems", the metaphor associating networks with organisms has lasted. To track down the paths of invention of the network, I argue that, as a technique, it cannot be separated from its representations as a "techno-imaginary" – a technique-network and technology-network –, and particularly not from its organistic metaphors.

The network was formalized and mathematised in the early nineteenth century, when it became a grid for interpreting space-time: a space-time matrix or rather a matrix of the "territory" it envelops like a new body. It became a "territorial network". Historian of techniques André Guillerme (1988: 8) points out that this modern meaning appeared at the beginning of the nineteenth century only, when the term was applied to basin hydrography (1802), to geology (1812), to the organization of fortifications on national territory (1821) and to the water distribution pipe system (1828). It was generalized as a result of the organization of a large system of communication channels and financial institutions in the Saint-Simonians' industrial "Manifesto" written in 1832, by one of their leaders Michel Chevalier (1806–1879). That is precisely when the double construction of the concept and the modern myth of the network occurred. A theory and a symbolic articulation of the network were to be the work of the Saint-Simonians.

The Cult of the Network Among Saint-Simonians and Proudhonians in the Nineteenth Century

For half a century from 1825 to 1875, particularly under the Second Empire, Saint-Simonian engineers and industrial actors worked towards developing railway networks, electrical telegraphy networks, and funding and training networks in France, Europe and the Arab countries. They theorized the industrial revolution and sought to socialize the major technical networks: this consisted in both conceiving of the socio-economic integration of the new networks, and devising modes of regulation, going so far as the "socialization of the means of production" which they suggested long before Marx, in *La Doctrine saint-simonienne* of 1829, under the impetus of one of their leaders, Saint-Amand Bazard (1791–1832). In order to carry out this socialization of the new territorial networks, particularly railways, the Saint-Simonians developed what was no less than a cult of the network, showing all the facets of the virtuous alliance that enabled an evolution from the communion of brothers to universal association and communication through networks. They enacted this communion in their church, staged the "association of the brothers" in their workshops and work seminars, and illustrated communication in their industrial and financial network policy. Communion was to proceed from the associated brothers' work applied to the entire planet, for the fertilization of nature with communication networks. Through such public interest work, the world could be reconfigured into an ideal organism composed of artificial networks that would transform it. Saint-Simonian religious practice consisted in creating an ideal artificial body, by drawing networks and superposing them onto the "natural" body of France and the Mediterranean, in other words its territory, to ensure the circulation of all flows in society. The object to be enveloped by the technical network was no longer just the organism or nature, but territory and society as a whole. Saint-Simonian engineers and entrepreneurs established themselves as the prophets and actors of this new technical, industrial and financial encircling.

Saint-Simon's doctrine was reformulated in modern terms of territorial networks primarily by the economist Michel Chevalier. In order to produce the modern idea of the territorial network, he had to fetishize the technical object to make it the symbol of "universal association". The territorial network could thus become the object of a cult through which the new technical network was equated with a radical change of society. This myth is still very much alive, as it is revived with every technical innovation, from the railway to electricity, IT or Internet and social networks. It conveys the belief that creating a new technical network amounts to triggering a change of society, economic mode of production, or even civilization. Michel Chevalier was the first to formulate this modern myth, in the early 1830s. Pierre-Joseph Proudhon (1809–1865), the father of anarchism, then reformulated it by creating a political cleavage within technical networks: depending on whether the network is centralized or not, the vision articulated will be either monarchical or revolutionary.

The Reification of the Concept of Network by Michel Chevalier

With the famous article-manifesto on the Saint-Simonians' industrial policy, *Le Système de la Méditerranée* (*The System of the Mediterranean*), published in the newspaper *Le Globe* on 12th February 1832 under the name of its editor Michel Chevalier, the network became the symbol of universal association. Following the schism of the Saint-Simonian Church in November 1831, this text translated the doctrine into a symbolism and cult of the network. The transition from the domination of men to the association of brothers could be made possible only by the development of communication networks, with communion and communication between East and West. With the network, the struggle between East and West could be "passed through" and "surpassed": it would unite the two, the flesh and the spirit, woman and man. East–West communion was identical in nature to that between the flesh and the spirit in the Christian religion. As another Saint-Simonian leader, Emile Barrault (1799–1869) declared: "Now that I have painted you a picture of the struggle and pacification of the East and the West in humanity, I can easily reveal to you, in each of you, these two worlds under the names of spirit and flesh, of thought and action, of intelligence and matter, struggling against each other and waiting for a law to harmonize them".[2] This fusion, a sort of Eucharist, is symbolically accomplished by the network which, for the believers of *New Christianity*, played the same role as Christ in traditional Christianity: a place of transubstantiation between body and spirit. With *The System of the Mediterranean*, Chevalier translated the schismatic split in the Saint-Simonian Church into action by placing the construction of communication networks at the center of their new cult.[3] If we are to appreciate the immense impact of this article, we need to consider it within the context of broader Saint-Simonian reflection at the beginning of 1832. It was the application of a "sermon" by Emile Barrault on East–West communication, delivered on 15th January 1832, which was essentially about struggle and the communion between spirit and flesh. This sermon required Barrault to "briefly outline their struggle in humanity between the peoples that have been its most energetic representatives, between the East and the West, followed by their impending reconciliation". Barrault posited an opposition as the starting point of his reasoning, an "eternal dualism", in other words a general contradiction, that between the East and the West. It then became a matter of knowing how to overcome this fundamental opposition, how to move from the struggle between

---

[2]Émile Barrault, Sermon of 15 January 1832, *Le Globe,* 16 January 1832.

[3]The first four articles, including *"Le Système de la Méditerranée"*, are titled "La paix est aujourd'hui la condition de l'émancipation des peuples" (Peace is now the condition of the emancipation of peoples), and signed by Michel Chevalier. They were published on 20 and 31 January, and 5 and 12 February 1832. They followed the publication of Émile Barrault's sermon in *Le Globe* on 16 January and inaugurated a series of Saint-Simonian propositions on the development of industrial policy.

two generic terms to their union, to harmony, and then to universal association. The Mediterranean, the historical locus of East-West confrontation, had to become their cradle of communion, through their envelopment by the communication networks that would allow the transition from domination to association. The encirclement of the Mediterranean by railways connecting the major harbors and by telegraphic networks provided the means to implement *New Christianity* (the title of Saint-Simon's last book), with a view to achieving communion between East and West. Barrault hoped to see this "new religion" prevail, to finally "unite in a solemn marriage the spirit and matter, science and industry, theory and practice, the East and the West, until then bound to struggle and antagonism! And what a moving spectacle humanity will present, when on the edges of the Mediterranean... Europe, Africa and Asia, as though on the edges of an immense and magnificent cup where they had made communion but by staining it with their blood, will now reach out with open arms of friendship and make peaceful communion together, and in this sublime harmony, will provide the symbol of the universal association we have just founded". Two weeks after the publication of Barrault's sermon, Chevalier started his series of articles presenting the project called *The System of the Mediterranean*, with which to connect the East and the West through a host of communication channels, and which prefigured "the universal association" by developing generalized circulation and international trade. In the first article, Chevalier added to Barrault's equivalences in relation to the East-West pair, with that of industry versus war: "Industry is eminently pacific. It instinctively rejects war. That which creates cannot reconcile with that which kills".[4] On that basis, Chevalier proposed "the main outlines of a plan" intended to "eternally secure a pacific future of prosperity and glory for the peoples of the world".[5] But he did not wish to simply demonstrate that peace is essential, he also sought to offer a "practical, implementable conclusion" and to provide means of action. Saint-Simonians saw the East-West conflict as the matrix of all social conflict, the most crucial of all. They wanted to see the Mediterranean transformed from a battlefield into a space of cooperation and association, a driver of universal peace. How could the Mediterranean evolve from a battleground to the "nuptial bed of the East and the West"? How could "the political consecration of harmony between matter and the spirit" be achieved at the same time?

"The Mediterranean", wrote Chevalier, "has been an arena, a closed field where, for three centuries, the East and the West have fought each other. The Mediterranean now ought to be a vast forum around all of which previously divided peoples will unite". From the arena to the forum, Chevalier made communication networks the instruments of an industrial and pacific construction, the technical matrices of the development of the Mediterranean basin.

---

[4]Michel Chevalier, "La paix est aujourd'hui la condition de l'émancipation des peuples", *Le Globe,* 30 January 1832.
[5]Ibid., *Le Globe,* 5 February 1832.

The Network as an Action Lever

The "general system" of the Mediterranean imagined by Chevalier makes each major port in the Mediterranean a place of interconnection of interlinking networks between the land, the sea and the inland waters. It even prioritizes the networks, into primary and secondary: "The port thus determined will serve as a pivot for a host of operations, the most crucial of which would be a railway. Going up the median valley, it would journey above or through waterways, to find another major valley. For the large river basins generally constitute the most natural industrial divisions, and all these partial systems tied together would constitute the general system. (...) All around the Mediterranean will thus be a primary network onto which secondary networks will be woven, especially so that the lines of communication converge towards the ports, which will serve as centers for each basin". There again, networks are connected to one another to create a system. Michel Chevalier described what should constitute the pivotal ports and associated networks, to serve Spain, Italy, Germany, Turkey, Russia, Asia and Africa, and thus painted "the delightful picture of what the old Continent would soon be". To this end, he deployed the full technical and symbolic wealth of the notion of communication network, even drawing on the metaphor of the body: "such a railway", he wrote with regard to Spain, "with all its branches... would be like a system of veins and arteries along which civilization in motion would awaken dozing Spain from its slumber, and connect its disjointed limbs". For the entire *The System of the Mediterranean*, he envisaged "a vast system of banks spreading a healthy chyle in all the veins of this body with raging activity, and countless joints". The artefact network brings the territory it envelops to life and fertilizes it, just as the natural network is meant to ensure the body's life. The technical network weaves itself into the territory and thus becomes a territorial network. "Such is our political plan", Chevalier concluded, "combined with the moral work designed by our supreme Father, of whom it is the material translation. It shall one day ensure the triumph of our faith".[6] In the early 1830s Chevalier's project, the material translation of the doctrine became the action programme of many Saint-Simonians. In these articles of the newspaper *Le Globe*, Saint-Simonian religious practice was asserted as an industrial cult and a political-financial communion around the fetishized communication networks. The network was seen as a link that could be both material and spiritual (here referred to as "immaterial"). This is what Michel Chevalier put forward: "Industry, leaving industrial actors aside, is comprised of production centers held together by a relatively material link, that is, by transport routes, and by a relatively spiritual link, that is, by banks. (...) There are such close relations between the bank network and the transport network, that if one of them is designed according to the best suited configuration for the exploitation of the planet, by virtue thereof the other sees its fundamental elements determined in the same manner". The modern notion of network was explicitly used for the first time by

---

[6]Ibid., *Le Globe,* 12 February 1832.

Chevalier, who thus distinguished between two families of technical networks – material, such as transport, and immaterial, such as the banking system – and at the same time emphasized their interdependence. This distinction was a cornerstone of modern thought on networks, which still associates the material infrastructure-network with an immaterial management, exploitation or financing network. But Chevalier took this further by specifying that communication networks had until then been the preserve of engineers alone, whereas their political significance was decisive, insofar as they contributed to achieving universal association: "Since those who have studied them [the means of communication] the most are engineers and do not claim to be anything else, the political and moral question has been neglected and the focus has been restricted to technical issues". In other words, the network was understood as both a technique creating ties – combining a material infrastructure and immaterial funding – and a political-moral operator serving as a system. Thus reified and fetishized, the network operates on two fronts: the one technical-financial and the other political-symbolic. It is more than a technique and an instrument of transition; it is the symbolic and practical operator of the Saint-Simonian religion, enabling the merging of East and West, of mind and body. The network is both a means of overcoming the original conflict and an end, as it definitively resolves the contradiction by creating pacifist universal association. Thanks to it, war is transformed into its opposite, universal association.

Networks thus become more than technical matrices built by engineers: they are symbols of social transformation, facilitating the transition from conflict to communion. To grasp their significance, their appearance as technical infrastructure needs to be overlooked. In other words, seeing the network as a technical object simultaneously amounts to effacing it, to reveal its truth in universal association. While the reification of the network is a first step towards its fetishization, the latter in turn reveals a symbol behind a thing. *The System of the Mediterranean* conceals networks as things (technical links) and reveals them as symbols (social links): "In the eyes of the men who believe that humanity is moving towards universal association, and who devote themselves to leading it there, railways appear in an entirely different light. The railways along which men and products can move around at a speed that would have been deemed mythical twenty years ago will remarkably multiply relations between peoples and between cities. In the material order, the railway is the most perfect symbol of universal association. The railways will change the conditions of human existence".[7] Chevalier saw the development of networks as a political revolution, turning a communication technique into a policy: "The large-scale introduction of railways on the continents and of boats on the seas will constitute not only an industrial revolution, but a political one too. Using these, and with the help of a few other modern discoveries, such as the telegraph, it will become easy to govern the major part of the continents surrounding the Mediterranean".[8] This assertion contains one of the main themes of the contemporary

---

[7]Michel Chevalier, *Le Globe,* 12 February 1832.
[8]Le Système de la Méditerranée, *Le Globe,* 12 February 1832.

symbolism of communication networks, which has become an ideology: the technical network fetishized as an instrument of social transformation. The technical part is equated to the totality of the social dimension. It follows from this that the technical network amounts to social change. Chevalier wrote that the day communication networks were developed, "an immense change will have occurred in the constitution of the world", for technical networks directly produced social change. Michel Chevalier even went so far as to describe the founding phases of network ideology: "Improving communication means working towards real, positive and practical freedom... it means creating equality and democracy. Perfected means of transportation have the effect of reducing distances, not only from one point to the other, but also from one class to the other" (Chevalier 1836, vol.II: 3). Technical communication networks inherently bear positive social change: the elimination of, or collaboration between, social classes. Communication networks mean democracy, association and equality. Conversely, the social issue (reducing the distance between classes) is pared down to a technical issue (creating communication networks). The engineer becomes the leading architect of social transformation. By reifying the network, Chevalier transformed a contradictory tension into a non-contradictory connection. Thanks to the network the contradiction is turned around or reversed into an association. The technical network enables communication, communion and democratization through the egalitarian movement of people. The geographical reduction of physical distance, even the interchangeability of places, owing to communication channels, results in the reduction of social distances, in other words democracy. In his Political Economy course at the Collège de France, in 1841–1842, Chevalier declared that "railways are democratic agents, in the legitimate and regular sense of the term. They put within the reach of all classes a means of transport that eliminates previously existing inequalities in the means of communication accessible to people" (Chevalier 1842: 378).

Michel Chevalier established a technocratic and liberal understanding of society which has been perpetuated in the contemporary ideology of the network: the technical network is now synonymous of democracy, movement, equality. Through the network, contradiction is eliminated, transformed into its opposite, the communion of opposites. That is why implementing the technical communication network in and of itself amounts to social change. Chevalier's liberal industrialism was founded solely on the virtue of the multiplication of communication networks to transform society. This theoretical position also supported a political position: it was not unrelated to the increased proximity between Michel Chevalier and the government, which was criticized by Barthélémy-Prosper Enfantin (1796–1864). Indeed, after turning down a venture in Egypt for an official mission in the United States, Chevalier publically announced his split from Saint-Simonianism.

In 1832 the word "network" in its modern sense appeared throughout Saint-Simonian engineers' writings, not only in Michel Chevalier's articles in *Le Globe*, but also in a collective volume published in September, *Vues politiques et pratiques sur les travaux publics en France* ("Political and practical views on public works in France"), authored by the engineers Lamé, Clapeyron and the Flachat brothers

(Lamé et al. 1832).⁹ These Saint-Simonian texts on the network answered "Father Enfantin's" request for Michel Chevalier, Stéphane Flachat and Henri Fournel to devise a plan on the "work particular to France" designed "as a first element of a general undertaking" to fulfil the project of "universal association".¹⁰ Stéphane Flachat surrounded himself with a group of Saint-Simonian engineers, including his brother Eugène, to write *Vues politiques et pratiques sur les travaux publics de France*. In this book, they discussed technical networks and the related financing and regulatory issues. It is an important work for two reasons: first, it often uses the term "network" in the modern sense (of a territorial technical network) and second, it constitutes the first systematic Saint-Simonian contribution to the elaboration of a theory of modes of network regulation. If we compare the meaning of the word "network" in two 1832 Saint-Simonian texts, namely Chevalier's article and the edited volume *Vues politiques et pratiques*, we see that both use the term in the modern sense of a technical communication system planning and developing a territory. The authors of the book thus noted that from the eighteenth century, by creating "a large network of royal roads", the State "covered the territory with a vast network of roads connecting the most remote regions of France to those where civilization, industry and agriculture were the most advanced" (Lamé et al. 1832; 27 and 33). In both texts, a "general communication system" operationally translated into the combination of several artificial networks. A system was therefore defined as a network of networks, following the organic model. For Michel Chevalier, the interweaving of networks (material and immaterial, primary and secondary) generated the general communication system. The same idea is found in *Vues politiques et pratiques*, through the combination of railways and canals, or of networks of different sizes: "Our general internal communications system must consist: (1) for primary networks, of wide canals and railroads for locomotive engines; (2) for secondary networks, of narrow canals and railways worked by horses and machines" (ibid.: 91). The combination of networks is the practical translation of the generalized communication system that Saint-Simonians called "universal association", and which Michel Chevalier defined as follows: "from the political point of view per se, universal association is the organization of a system of industrial works that embraces the entire world".¹¹ The symbolism of universal association is translated into the implementation and interconnection of technical networks enveloping the world. From this perspective, the interweaving of networks leads to the formation of universal association: the network must be "put

---

⁹Gabriel Lamé (1795–1870), an 1814 graduate of the École Polytechnique, became a physics professor at this school. Émile Clapeyron (1799–1864), an 1816 graduate of the École Polytechnique, took part in the construction of the Paris-Versailles-Saint-Germain railway and was elected to the Corps Législatif in 1868.

¹⁰This is what Michel Chevalier reported in his article in *Le Globe* on 30 March 1832, entitled "Politique d'association, politique de déplacement" ("Association Policy, Movement Policy") – cited in the pamphlet *Politique industrielle et Système de la Méditerranée*, June 1832. Rue Monsigny, n°6. Paris (Paris – Fonds Enfantin Bibliothèque de l'Arsenal, FE 957, pp. 29–39).

¹¹Michel Chevalier, *Le Globe*, 30 March 1830.

together" by "interconnecting water transport, internally and externally", wrote the authors of *Vues politiques et pratiques* (Lamé et al. 1832: 102). The interconnected networks constitute a sort of fabric which envelops territory and society. Covering the planet with networks and thereby fertilizing the body of the Earth-woman was the modern myth founded by Saint-Simonian engineers, and which is still pursued by technical network development policies to this day. The "System" built by Saint-Simonian engineers likens the feminine Earth encircled with artificial networks to a living organism, a network of natural networks. The engineers built the ideal social system of "universal association" from a complex combination of differentiated artificial networks, primary and secondary, material and spiritual, railway networks, road networks, canal and telegraphy networks, etc. The cult of the Saint-Simonian religion is expressed rationally, in the construction of networks, for engineers are established as demiurges capable of computing and creating a social system by combining networking artefacts. If the network is the elementary structure of a system, a social system can be planned through a combination of networks: this is the plan of action of *The System of the Mediterranean*. A complex system can be composed by combining networks of very different natures (banking and communication), sizes (primary and secondary) or types (canals-railways-telegraphy). It thus becomes possible for engineers to imagine and construct an ideal social body, through the combination of networks interconnected on the model of the human body. Nineteenth century Saint-Simonian engineers' know-how was thus extended and applied to the treatment of the social body, not least by sociologist-engineer Herbert Spencer later on. The demiurge engineers, capable of creating the networks of an ideal social body, were the best suited to define the conditions of implementation and exploitation, including the modes of regulation. The authors of *Vues politiques et pratiques sur les travaux publics en France* argued that the regulation of networks could be thought of in terms of a trilogy of hypotheses, which I will refer to as "the thesis of the three theses". The management of a network can be entrusted either to the State or to private companies, or else be performed through a mixed solution, such as concession: "Three systems can be envisaged for the implementation of public projects: (1) The government may be solely responsible for implementation with funds obtained from taxes or loans. (2) Implementation may be left entirely to companies' speculation; it is then their responsibility to determine which works promise sufficient returns to justify the spending they have assessed, and then to collect the funds, manage construction, oversee maintenance, and see to improvements. (3) Implementation may be entrusted to companies, subsidized and monitored by the government. The companies execute the terms they agreed to, in which the main details of the projects are set out. They comply with certain conditions for maintenance, improvements, reduced rates, etc." (Lamé et al. 1832: 256–257). The economic and political regulation of the network advocated in this book corresponds to its symbolic-political function. Because the network symbolically performs the transition from a military and state society to an industrial and entrepreneurial society, its economic regulation can only be an intermediary between State and enterprise. To adequately fulfil its "transition" function, the network requires "mixed" regulation, between State and enterprise, public and private. In his *Cours d'Economie Politique* at the Collège de France, Michel Chevalier extensively developed this thesis of "mixed industrialism", a mix of interventionism and liberalism. The "thesis of the three theses" seems consubstantial with the notion of network: it is the equivalent, for the issue of regulation, of the symbolism of the network contrasting opposite images to better reunify them with the idea of association or communion of opposites.

In 1830, three Saint-Simonian leaders, Prosper Enfantin, Saint-Amand Bazard and Michel Chevalier came together, symbolizing three possible paths for the potential development of the Saint-Simonian theory of the network. In November 1831, the first (and major) schism of the school took place, between Bazard on the one hand and Enfantin-Chevalier on the other: the interpretation of the Saint-Simonian theory as a tool for social transformation was excluded in favor of a religion based on the reformist cult of networks, advocated by Bazard. In March 1833, a second schism took place, between Enfantin and Chevalier: the former emphasized the religious aspect of the cult of the network, while the latter prioritized a liberal and technocratic understanding of the development of networks. In October 1833, Enfantin left for Egypt to participate in building the Suez Canal and to implement the Saint-Simonian symbolism of networks, for the achievement of universal association. At the same time, Chevalier had gone to the United States to study networks, and was advocating a political economy of communication networks as an end in itself. In the wake of these "schisms", the theoretical unity of the network concept was shattered. Within the Saint-Simonian movement itself, it was split into at least three separate understandings, which can be simplified to the extreme as: (1) a policy of social transformation that uses the network concept to think about the transition towards a future society, with Bazard; (2) a religion of universal communication carried out by the networks fertilizing Mother-Earth, with Enfantin; (3) a liberal and technocratic political economy of communication networks, with Chevalier. Later approaches to the notion of network, which observed its multidimensionality, even its indeterminacy, merely brought together and tinkered with scattered pieces of this Saint-Simonian "fragmentation".

Once the demarcation line was drawn within the movement, most Saint-Simonians, starting with "Father Enfantin", applied themselves to creating technical and financial networks, which then constituted Saint-Simonian religious practice. These Saint-Simonians were to form "a saint militia... an army, under the banner of universal association", as one of them, Charles Duveyrier (1803–1866), wrote as he urged the School to take action: "Words are therefore no longer enough, we need facts; we need to move from speech to action, from the programme to the enterprise".[12] After the failure of their retreat at Ménilmontant, the Saint-Simonians scattered. However, many of them, particularly Enfantin's relatives, devoted themselves to the creation of networks for the movement of money, knowledge and communication, until the end of the Second Empire. In 1858, Enfantin took stock of the Saint-Simonian industrial and financial promotion of the development of

---

[12]Charles Duveyrier, Politique industrielle, *Le Globe,* 21 February 1832.

railways, the electrical telegraph and banking and financial networks, and reflected on Chevalier's industrial *Manifesto*: "This is why Michel wrote the *System of the Mediterranean* where he outlined the plan of this gigantic building project that no one but us was considering at the time, which cost billions and is now almost finished" (Enfantin 1877 in Saint-Simon and Enfantin 1865–1878, vol. 46: 211). Enfantin considered his work done; in his words, "we have embraced the Earth with our networks of railways, gold, silver, electricity! Spread the spirit of God and the education of humankind through these new channels, of which you are partly the creators and masters" (Enfantin, *Le Crédit intellectuel*, 1866 quoted by Pinet 1898: 165–166). This declaration neatly sums up the intention behind the Saint Simonians' action: communication networks were created as a religious practice to "embrace the planet". It was a real act of love of the Earth fertilized by the network, an envelopment of society by technical networks. All Saint-Simonian projects were embedded in this religion of universal association, seeking to develop a generalized circulation of flows across the world, reflecting the perfection of the network.

Saint Simonians were the first to think about railways in terms of networks and to see a political and social revolution therein. Others followed, such as anarchist thinkers Proudhon and Kropotkin, who went so far as to introduce a political cleavage within the mode of development and the architecture of communication networks, between the advocates of full state centralization and those of decentralizing equality. Proudhon limited the revolutionary scope of networks as envisaged by Enfantin and Chevalier: for him, the network could bring about a social revolution only under certain political conditions of organization and regulation. The Saint-Simonian myth of social transformation achieved automatically by the development of a new communication network was reformulated by Proudhon, who saw the very architecture of the technical network as a societal choice.

A Type of Society Embedded in the Structure of the Network

With Saint-Simonianism, the network enters society and becomes socialized. With Proudhon, it is society that enters the network. The technical network is not only a means to envelop territory and society, for, according to Proudhon and Kropotkin, the choice of a social system is nested within the internal structure of technical networks. Proudhon thus saw a mode of social organization in the structure of technical networks. A centralized network means a centralized society and vice versa. Proudhon applied this analysis to railways, and Kropotkin later used it for electrical networks. It implies that the technical network and society define each other through the similitude of their structures. The process of fetishization of the network set in motion by Chevalier, which consisted in taking the particular (the technical network) as the global (universal association, the change of society), was extended by Proudhon. Indeed, he brought fetishization into the very architecture of the technical network: a part of the network, its structure or its "framework", amounts to a type of socio-political organization.

Proudhon was of the mind that exchanges and flows in society had to be multiplied in order to increase individuals' freedom and to improve social dynamics. Proudhon and Chevalier shared the same starting point: the analogy between the human body and the social body: "Just as blood circulation is the mother and driving function of the human body, so too the circulation of products is the mother and driving function of the social body", wrote Proudhon (1851: 201). The revolutionary political programme of June 1848 was grounded in this principle: "All the ills afflicting the social body can be related to the cessation or to a disruption of the circulatory function. The circulation is nil. There is a crisis" (1848: 140). The railway and waterway transport networks are the very symbol of this circulatory mobility. In terms very close to those of Michel Chevalier, Proudhon wrote in 1845: "Railways remove the intervals, make people present with one another everywhere... railways, due to the nature of their service and to their phenomenal development, affect everything and determine everything". The railway "erases and levels all inequalities of position and climate" (1868a: 264 and 297). But this positive effect of the network is perverted by political centralization and economic monopoly. In 1855, following the great law of 1842 which organized the railway lines in France radially from Paris, Proudhon wrote *Des réformes à opérer dans l'exploitation des chemins de fer* (1868b, vol. XII), to speak out against the State monopoly of railways. He advocated a mixed system in which the State funds infrastructures and leaves the management in the hands of private companies. He thus drew on the "thesis of the three theses" formulated by Lamé, Clapeyron and Flachat. Proudhon stressed the importance of the images and representations associated with a technical network, particularly in its early days, and had already adopted a critical stance with regard to these discourses: "After 30 years of existence, from a political economy perspective the railway is still a myth", he commented in his preface. "The public itself, after indulging in the most fanciful hopes, was then overcome with wariness and tormented by the most insane imaginings. (...) Not to mention, alongside the chattering of the sages and quackery, the alternately apologetic and accusatory clamor of the subordinate interests which, depending on the sentiment animating them, take a stand for or against railways, curse them or laud them" (ibid.: 2 and 4), he wrote, with emphases that we still find articulated today, "for or against" the Internet. He added that the railway "is used as a theme by a new kind of agitator", and recalled that "an archbishop, in a sermon at Lent, denounced the railway before his pious flock, as signs of revenge from the sky for the incredulity of men. An even more fanatic author, announcing the arrival of the antichrist, warned that the electric telegraph and the locomotive were symbols of its cursed power. Democracy, on the other hand, salutes railways as vehicles of equality, more effective than those of 1793" (ibid.: 5). Proudhon was the first to criticize the ambivalent social representations of the technical network: for some, it was a curse, the symbol of power, for others, the symbol of equality. Having criticized the symbolic dimension of networks, he in turn took a stand to shift the terms of the political-symbolic confrontation. He replaced the antichrist-democracy opposition he denounced with the "more real" contradiction between the centralization of power exercised on the network and another form of management based on "federating unity": "Waterways, primitively provided by nature, left the Gallic territory divided into as many commercial regions, independent of one another, as there were drainage areas. Roads were intended to unite these separate regions; canals had no other destination. All of these channels put together therefore form a system of general equality, a sort of federating unity. By decreeing railways, the 1842 law seems to have wanted to change this whole tradition, which is not only commercial and political, but also pertains to transport. Instead of continuing the work that canals and roads had started so well, the law followed the impetus of the monarchical idea that sees Paris as the Queen of Gaul, and each province as a fiefdom that is tributary to the capital. All our railways, like beams, start at the center of the government". Proudhon was the first to contrast two categories of technical networks comparable to two political visions applied to reticular artefacts: "On the chequered network, a federating and egalitarian network of land roads and waterways, has thus been superimposed the monarchical and centralizing network of railways, which tends to subordinate the départements to the capital" (ibid.: 97–98). Proudhon contrasted the figure of the chequered network, seen as "natural" and egalitarian, with that of the monarchical and centralized "artificial" network, characteristic of "the princely, governmental concept of the radial network". The chequered network contrasts with the radial network: the forms of power are reinvested into the reticular technical architecture. Legrand's radial organization of the railways,[13] with Paris at the center, like the system of optical telegraph lines, was a fine illustration of Jacobin power and economic monopoly. The chequered organization is the opposite of the radial organization: in other words, a political choice is embedded in the very architecture of the technical network. In substance, Proudhon argued that it is pointless supporting mythical discourses "outside of" technique (antichrist *versus* democracy of 1793); the regulation and organization of the network should be criticized from "the inside". The structure of a technical network conveys a choice of economic policy: this Proudhonian assertion has been abundantly drawn upon to this day, articulated in the form of equivalence between the structure of a technical network and the organization of a society. According to Proudhon, the railway inherently produces a beneficial revolution of human relations, by multiplying interaction and removing intermediaries: "by virtue of the consistency and regularity of their service, further aided by telegraphic correspondence, railways have the effect of bringing the producer and consumer into direct contact, irrespective of the distance between them, and consequently of removing intermediaries" (Proudhon 1855: 293). The network is an "admirable" figure in anarchist thinking, for it defines pure transition or flow, a direct relationship that cannot be institutionalized. Proudhon explicitly saw duality in the railway: he identified it as both a circulation technique and the symbol of an economic policy.

---

[13] Alexis Legrand (1791–1848), general manager of the department of civil engineering, outlined the plans of the first French railway network centralised in Paris, hence the name "Legrand's star".

His criticism was concerned with the discrepancy between the technique and what it symbolized (freedom and equality). By reuniting the technical and social forms of organization, in other words by disseminating a technique democratically, in accordance with its potential, the symbol will disappear: "since the railway has communicated its eminent qualities... to the whole social order through a sort of magnetization, it is reasonable to predict that one day the railway as a symbol will become worn-out" (ibid.: 366). Proudhon thus saw the technological fetishism promoted by Saint-Simonians as simply a moment in the development of a network; one that faded with the social dissemination of the use of techniques.

Proudhon's opposition between centralization and the federating structure with regard to railways was used by Peter Kropotkin (1842–1921) in relation to electricity: the electrical network affords the possibility of decentralized structures and organizations. One of the effects of the "reticular revolution" is its capacity to transform organizations, to shift from centralized and hierarchical structures to small entities associated through networks. In his book *Fields, Factories and Workshops*, Kropotkin (1910 [1898]) argued that electricity provides new impetus to small industries, and used the example of businesses in the Monts des Lyonnais and certain rural regions of England. He maintained that small industry develops alongside large centralized industries, "where waterfalls have been exploited to obtain electrical power in villages, and where machines were used in large cities to produce electrical light during the night... the small industries are experiencing a new expansion" (ibid.: 281). He saw this as a means of transition towards free federations of groups of producers and consumers. Thanks to electricity, small industries can develop, including by working during the night in the cities, and work from home, which is less tiring than in the large factories, becomes possible: "Far from disappearing, on the contrary these small industries tend to develop, especially since in certain large cities, like Manchester, electricity has afforded a cheap driving force, in the exact measure required in a given moment" (ibid.: 253). Kropotkin gave multiple examples to show that the electrical network allowed for small production units, on a human and family scale, "the small establishments where manufacture can take place in the best conditions", as opposed to "monstrous factories", or "large factories": the network makes decentralization and even self-management possible. Not only does electricity ensure productivity, for there is constant lighting to work, it makes it possible to remain in one's usual environment, thus avoiding the uprooting of workers with their forced migration to the big cities.

The Technological Utopia of the Network

From the Saint-Simonians to Proudhon and Kropotkin, that is to say, in a century during which railway, telegraphy and electricity networks developed, the modern notion of network unfolded in all its complexity, as a myth and as a territorial technical matrix.

To its credit, the fetishization of technical networks facilitates their social dissemination, by transforming them into symbols of a policy of generalized exchange, that of "universal association". But according to Proudhon, this over-symbolization conceals a more concrete issue: the economic policies of network regulation inherent in their very mode of organization. By identifying political choices within the architecture of networks, Proudhon sought to drain the symbolic excesses of some Saint-Simonians, particularly Chevalier. The Saint-Simonians had rid the social utopia of its burden, by transferring the promise of social change to the technical network, the railway, a symbol of "universal association". This transmutation of the Saint-Simonian "social semi-utopia", to use philosopher Raymond Ruyer's (1950) expression, into a full-blown technological utopia, was achieved through the fetishization of technical networks. Proudhon therefore reacted to this excessive symbolization of networks that legitimated the theoretical-political stance of the liberal wing of the Saint-Simonians close to Napoleon III, by re-embedding the political into the organization of networks. In concrete terms, the technical network is admittedly more of an agent of social transformation in its internal architecture than in the images it conveys. With the twofold Saint-Simonian and Proudhonian intervention, the terms of the debate on networks were lastingly set and were to become real ideological "markers". First, the concept of network deteriorated into a "technology of the mind" and second, the modern myth of transformation achieved through the technical network, even in its architecture, prevailed and was repeated with the emergence of each reticular innovation. This recurrent modern myth announced a new social and economic revolution with the birth of electricity, the telephone, the computer, CTI, cable, satellite and the Internet.

This myth, developed around 1830, was reactivated with the appearance of each new technical network: electricity, which Lenin claimed defined socialism by associating it with the "power of the Soviets", and the telephone, followed by Internet, considered as the "nervous systems" of society. In the mid-1990s, US vice-president Al Gore declared to the international community: "the President of the United States and I believe that an essential prerequisite to sustainable development, for all members of the human family, is the creation of this network of networks [Global Information Infrastructure – GII]. [It] will circle the globe with information superhighways on which all people can travel. (...) The distributed intelligence of the GII will spread participatory democracy... I see a new Athenian Age of democracy forged in the fora the GII will create"! Commenting on the Arab Spring in 2011, Hillary Clinton asserted that "Internet is freedom!", once again turning a technique into a symbol.

The intense period of invention of the early nineteenth century has left us with the legacy of a "technology of the mind" and a "techno-messianism", as well as their combinations; in short, an ideology of the network that we here call a *retiology*, which is readily presented in the form of a utopia.

Retiology *or Network Ideology*

Contemporary *retiology* combines two deteriorated and worn-out components: first, a concept that has become a technology of the mind and second, a symbolic or techno-utopian (or "techno-messianian") operation through the fetishization of reticular techniques. First of all, what do we mean by "technology of the mind"? This is a canonical reasoning process through which engineers and industrialists theorize their design, construction and regulation practices in relation to territorial technical macro-networks.

Let me also clarify what is meant by "techno-utopia" or "technological utopia", the second component of *retiology*, which comes in two forms: either "techno-messianism" (positive vision), or "techno-catastrophism (negative version). The Saint-Simonians and the Proudhonians were the first militant-manufacturers of the images associated with reticular techniques. The technical network and even its architecture then carried the images of a new society and of "universal association". Finally, contemporary *retiology* is comprised of the remnants of the dilapidation of reticular symbolism, reduced to imagery and narratives, which support the practices and the promotion of the technique.

The ideological triumph of the network relates to its theoretical vacuity and its loss of symbolic references. This *retiology* combines a concept reified into a technology and a utopia embedded in the fetishized technique, so as to celebrate each reticular innovation as an extravaganza announcing a "new world".

With the proliferation of technical networks since the end of the nineteenth century, the modern mythical narrative of social transformation through the network and its architecture has been reactivated and reviewed with each innovation of the reticulated techniques, from electricity to the Internet. This recurrent action to resuscitate a symbolicity of the reticulated has been officiated by mainly engineers and industrialists, who spread the webs of networks over the entire planet and throughout society. Engineers legitimize and socialize the artificial networks they design, using organic images of the reticulated. Engineer-sociologists envelop their technological productions with reticulated metaphors borrowed from the human body, until body and network, brain and computer, again become one and the same. Engineers endeavor to liken artificial networks to a living body, to associate them with corporal images, particularly pertaining to the brain, and thus to naturalize techniques. The electricity, electronic or telecommunication networks are even claimed to be "intelligent", to constitute a sort of "collective brain" or "global brain" which artificially embodies the Galenic metaphor (Galen's *rete mirabili*) on a global scale. The Saint-Simonian image of the social transformation effected by the technical network and embedded in its architecture by the Proudhonians has prevailed as a great modern myth, a narrative that has been vulgarized not only by engineers, industrialists and "futurologists", but also by certain sociologists. Chevalier and Proudhon's idea of a political and social structure inserted into the technical network, which acts as a modern "hidden God", explains this recurrent reactivation. The network becomes a lever for political and social transformation, perceived and used in all organizations to make them evolve. Just as the Enlightenment thinkers, particularly in Diderot and d'Alembert's Encyclopaedia, identified hidden networks everywhere in organisms and Nature, so too contemporary ideology decrypts them in organizations and territories.

This narrative, which is mobilized with every reticular technical innovation, always draws on corporal metaphors, particularly comparing the technical network with the nervous system and the network with the brain, so that it may prevail as a new figure of power or counter power, in organizations and in society. Engineer-sociologists and industrialists have used and worn out this figure, drawn from Saint-Simonianism, recurrently presenting each "new" technical network as a means of transforming society, the economy and organizations. This deterioration of the Saint-Simonian vision into "techno-messianism" comes in the form of a "techno-utopia". The techno-utopia of the technical network has been repeated with a few invariants, from electricity to the Internet. As for the concept of network, contemporary discourses have reduced it to a "technology of the mind", a frequently useful reasoning process, the content of which is however limited to describing relations or interconnections between elements of a fragmented whole.

Across the diversity of reticular practices, the network ideology eternally promises social change or, more simply, the creation of movement and mobility through "connection" prosthetics. The network now catches everything and anything in its nets. It has become an ideology, a *retiology* that recycles the symbolic images it has held, particularly the promise of a transition towards the future. The technical macro-networks, great contemporary technological and industrial undertakings, are the modern "cathedrals of the celebration of passage". They no longer reach towards a celestial beyond, but dramatize alternately the transition towards a better world to come and the continual setting in motion of the present.

The Reticular Techno-Utopia

The mythical narratives of the network are repeated to saturation point by engineers and industrialists to promote their innovations. The techno-utopia of the technical network again takes up two main themes: first, the ancient narrative of the inter-world between body and technique, introduced by Galen, which draws a parallel between the body, particularly the brain, and the network, and second the modern narrative stemming from Saint-Simonianism and Proudhonism, which sees the technical network as a lever, if not an identifier, of politics. The link between the body and the network is maintained, but the other way around, as it is now the body, particularly the brain, that serves as a model for conceptualizing the artificial network. As for the mythical Saint-Simonian narrative of social change effected by the technical transformation of networks, it is repeated with every technical innovation, reaching new heights with the Internet, which signaled the beginning of a "new Age", of a "new economy", and of the "network society" which Manuel Castells prophesied. The network-manufacturing engineers-sociologists use the images and metaphors of reticulated techniques to socialize their innovations and to take them out of their laboratories, so as to project them into society. The Saint-Simonian vulgate produced by engineers from the *École polytechnique* paved the way for the fetishization of the network with railways, and for the formulation of a "techno-utopia" of the network based on a few invariants or "markers". Like any mythical narrative, this techno-utopia relies on recurrent apologetic discourses (or terrifying discourses, which amount to the same thing, "the other way around") articulated at the time of the emergence of electricity networks at the end of the nineteenth century, and then of telecommunication and finally IT networks. These discourses always present the new network with reference to the organistic metaphor and to the technical-political utopia of social transformation: the new network will be "alive" and "revolutionary". Such dramatization seems necessary for the promotion of each technical reticular innovation. To develop industrially and culturally, the technical network must become a "technical system", as understood by Bertrand Gille (1978), even a "technical macro-system", as defined by Thomas Hugues (1983) and Alain Gras (1997), combining technical networks and power systems to form a whole.

The Six Markers or Invariants of Reticular Techno-Utopia

By the mid-nineteenth century, the mythical narrative of the network had essentially taken shape with the development of railways. The long history of the reticular imaginary that defined it as an inter-world between organisms and techniques was supplemented by the Saint-Simonian and Proudhonian contributions. These fictions were characterized by the theme of socio-technical "revolution" through the fragmentation of the existing society and the promise of a new one. They were also steeped in the metaphor of the brain or nervous system, stemming from the medicine of Galen, reworked by Descartes – who both likened the brain to a network or net –, and applied to new technical networks. Since it had the same logical functioning and a similar material architecture, the technical network was seen to be to society or the planet, what the brain and the nervous system were to the human body, that is, the regulatory organs. I suggest a selection of six "markers" to outline the contemporary mythical narrative of the network, shaped by the memory of the multiple contributions that made the Saint-Simonian operation possible, before expanding into a techno-utopia. They constitute the vulgate or *doxa* of discourse on the network, disseminated and rearticulated by engineer-sociologists with every innovation. These markers can be thought of as the scoria of the long work performed on the idea and image of the network, particularly during the first half of the nineteenth century.

1. The first, most powerful and oldest marker was set by Galen and then rearticulated by Ambroise Paré and Descartes. It associates the network and the body, and in particular likens the brain to the network. From the nineteenth

century this analogy was reversed, as the focus was then not so much on explaining the body through the network, as on legitimating the artificial network with the image of the natural organism. This marker can be labelled "biotechnological" or "Galenic-Cartesian" (Galen/Descartes). Just as the textile network had allowed for the rationalization of the organism, so the organism was used to naturalize the artificial network. From the nineteenth century onwards, images were traded between artificial and natural networks. The naturalization of the technical network helped it to become embedded in society by presenting it as necessary for the renewal of the social body. The network is comparable to a human organism, or to one of its parts (bloodstream or nervous system). As in the body, networks interlink and multiply to ensure fluidity and social flow. The modern artificial network draws its rationality, its regulatory model and its naturalization from the archaic figure of the body-organism. It gives itself a body to be socially integrated. The purpose of this metaphorical discourse of the organism, essentially articulated by engineers, is to convince the political sphere, industry, the market, users, etc. The crucial advantage of the model of the organism is that it immediately, "naturally" offers a mode of regulation to the new artificial network, before it becomes a "technical macro-network" over the long term. It provides the technical network with cohesion, harmony and expressivity: the body or organism-network is a model. Each organ, particularly the brain, or the body as a whole, can be taken as a model of rationality to think about and promote the new technical network. Moreover, the model of the organism-network naturalizes and acclimatizes the new technique, even makes it "user-friendly" and articulates it with the social, by tying in the parts with the whole. The technical network is thought of as the nervous system or bloodstream, or as the brain which regulates society as a whole. The corporal metaphor illustrates the technical-social shift announced by the emergence of the new technical network.

2. The second marker of the network is not so much Cartesian as Leibnizian; it signals that the network can be formalized. The network follows a logic, an order, even a "graphical reason".[14] It can always be drawn, in the form of a graph associated with a matrix. The functioning logic of the network is embedded in the outline of its structure, in its architecture. As Jacques Bertin highlighted in *Semiology of Graphics* (Bertin 2010), the graph offers a "rational image", which differs however from the figurative or mathematical image; it is "a language for the eye".[15] The network is represented by links (or correspondences) drawn up between places (or elements) on a map. Unlike the diagram, the network graph seeks "the most effective and simplest image possible". In the diagram, a meaning is attributed to the dimensions of the map before placing correspondences, whereas the network aims towards "the disposition that offers the least crossovers or the simplest figure". In other words, it seeks a meaningful, simple

---

[14]Based on the title of the book by Jacques Goody (1977).

[15]See in particular p. 269 and the following pages on networks.

and effective order, but without prior meaning: only the presence of elements and links matters.
3. The third Saint-Simonian marker stems from Michel Chevalier's *Manifesto*. The technical network always announces a technical and therefore social revolution, through the fragmentation of the existing social structure and the promise of modernity in the future. The network transforms society: electricity, IT, telecommunications, transport or the Internet produce "post" societies (post-modern, post-industrial, etc.). With the proliferation of technical networks since the end of the nineteenth century, the utopian narrative of social transformation accomplished by way of the network has been revived and modified with each new networked innovation. This recurrent enactment of the reticulated has been officiated primarily by engineers, who spread the webs of their networks across the planet and in all spheres of society. It is no longer a matter of enveloping Nature or the body in technical networks, as it was during the Enlightenment, but society as a whole, now and in the future. The network is seen as inherently "revolutionary": from Lenin's "soviets of electricity" to US vice-president Al Gore's "highways of information", the new networks are claimed to provide democracy, transparency, freedom and equality.
4. The fourth marker is again Saint-Simonian and formulated in Chevalier's *Manifesto*. Networks contribute to peace, prosperity and universal association, as they artificially cover the Earth. The network, like a net, is thrown over the planet and the society it envelops, and even becomes its structure, both visible and invisible. The city will become a "smart city", the planet will be "relational" and society will be "networked". Each individual, activity or object must be "interconnected" with and defined through networks, starting with atomized subjects. The relationship with territory is modified. Shorter time, reduced distance, greater speed: the network "brings nearer" and modifies, or even removes, territory. It becomes a tool to plan and develop the territory, starting with urban space.
5. The fifth marker, also Saint-Simonian and rooted in Chevalier's *Manifesto*, presents the network as a bearer of prosperity, progress in new activities, the multiplication of new services, a "new economy", etc. The network is an answer to crisis, by ensuring economic development and prosperity, and the technical network is a major figure of the great modern myth of Progress. As such, it bears the promise of new occupations, and a new cycle of growth. It even allows for the definition of an economic policy (for example, today's so-called "digital" policies), by moving the object of the traditional policy onto a technical, even technicist plane, entrusted to industrialists, experts and engineers.
6. The sixth marker is Proudhonian and libertarian. A choice of society or policy is embedded within the very architecture of the network. This complements the Leibnizian marker, which outlines the content of a form. Its graph either reveals a monarchical, Jacobin, centralizing policy, or the opposite. Its architecture conveys organizational choices regarding the State, businesses, verticality versus horizontality, centralization versus decentralization, etc. Proudhon and Kropotkin were the first to clarify the nature of this network revolution: networks

can decentralize and become a means of fighting concentration, they can even become a self-management tool. A policy can be read and "seen" in the architecture of the network. There again, the network shifts the political and embeds it in technical choices. To re-examine the Proudhonian marker, Langdon Winner (1986) interpreted the concept of "affordance" borrowed from engineering by Barry Wellman, to study the "political properties" of technologies.

To sum up these six markers of the techno-utopia of networks, we can say that the network is comparable to an organism, or to one of its parts – the brain – which gives it cohesion (Galenic-Cartesian marker); it reveals the forms of a graphical rationality (Leibnizian marker); the network bears the promise of a dual revolution, both technical and social (1st Saint-Simonian marker); it brings progress and prosperity, and provides a new political answer to crises (2nd Saint-Simonian marker); the network conveys universalism, by reducing distances and time-scales (3rd Saint-Simonian marker); finally, the network embeds and reveals social and political forms of organization in its architecture (Proudhonian marker).

The three Saint-Simonian markers complement one another. They assign to the technical network the outlines of a new social system, in accordance with Chevalier's 1832 *Manifesto*. The network is as much a symbol as it is a technical infrastructure. It is a means of transition to escape a crisis situation and achieve a state of progress, peace and prosperity. The Leibnizian and Proudhonian markers also complement each other as they reveal meaning in a form and in a reticular architecture. Finally, the Galenic-Cartesian marker is found throughout the history of the network, as it binds the network to the corporal metaphor, drawing alternately on the reticular model to rationalize the organism, and on the corporal model to naturalize and regulate the technical network. There are therefore three dimensions to the techno-utopian discourse on the network, which make its imaginary so powerful: a temporality of transition, an organic spatiality, and a graphical rationality. This "retiological" discourse is found in each of these three main dimensions:

- The network manages the social transition from a state of crisis to another state characterized by prosperity, progress, democracy and modernity;
- It naturalizes the technical network by way of metonymy, by relating the planet and the organism, or one of its parts, such as the nervous system or the brain;
- Finally, it delivers knowledge in what it shows, in a visible rationality, in a drawing.

Since the mid-nineteenth century, the technical network discourse has offered a linear temporality of social transition, a natural organic regulatory spatiality and a rationality that can be read in reticulated forms. It is as though the techno-utopia of the network offered three levels of interpretation: forms, flows and regulation. This triptych, comprised of readable forms, flows of transition and organistic regulation, has made the strength of the techno-utopia of the network and structured the recurrent discourses that go hand in hand with reticulated innovations, from the railway to the Internet and the NBIC (Nanoscience, Biotechnology, Information Technology and Cognitive Science).

The Bio-social View of the Technical Network

The techno-utopia of the network describes differentiated network architectures, announces a social revolution enabled by the new technical network, and reactivates the metaphor of the organism-network to naturalize and regulate networks. The techno-organistic fiction, particularly the image of the nervous system or of the brain, contributes to socializing technical innovation by transforming it into a techno-political utopia. Social change is thus thought to stem from the technical network, interpreted as a hybrid being that is both artificial and natural, an organism-network. Since Antiquity, doctor-philosophers had thought about and shed light on the human body through the image of the net and weaving in order to understand its fundamental principles – to the point of seeing them, by the end of the eighteenth century, as one and the same thing. The modern engineer has reversed this logic, to use reticulated images inspired by observed or imagined network effects in the human body and, by way of metonymy, in the nervous system and the brain. This reversal is used to promote technical innovation and emphasize its transformative social impact. It is as though, in the engineer's mind, "without a body the technical network could not live", just as, without the model of the network, the doctor could not interpret the body. The engineer endeavors to give the network a body. The first to systematize it, after Enfantin, was Herbert Spencer (1820–1903), a former railway engineer turned sociologist. He formulated the fiction of the organism-network. Drawing on Lamarckian and Darwinian theories, his *Principles of Sociology* (published in three volumes from 1877 to 1896) merged the social and biological functions, in a "social evolutionism". He referred to a "determinable order" and contrasted the "fighting and predator" system of past societies with the "industrial regime" of current societies: "the contrast between the fighting and industrial types hinges on replacing the belief that individuals exist for the benefit of the State with the belief that the State exists for the benefit of individuals". In fact, Spencer's theory is based on the Saint-Simonian opposition established by Enfantin, between an "organic" or industrial society and a "military" society, which he takes to the extreme with an organicist view of the social.[16] Spencer distinguished between three "organ apparatuses": producer, distributor and regulator. In the human body, the regulatory organ is the nervous system, which disseminates information in the organism: its equivalent in society is all the means

---

[16]"If the organisation consists of a construction of the whole such that it allows its parts to fulfil actions interconnected through mutual dependence, the less advanced the organisation is, the most interdependent the parts must be; while of the contrary, when the organisation is advanced, the parts' dependence is overall disastrous. This is something which is as true of the individual organism as of the social organism" (Spencer 1883-1890, vol. 2: 53–54).

of communication, whether postal services, the telegraph or press agencies. This view provided the first explicit formulation of the image of the network as a social nervous system: "The only telegraph wire", Spencer wrote, "covering all the branches of the railway system is the wire that stops or spurs its traffic, just as the nerve covering the full length of an artery is the vasomotor nerve that regulates its circulation. (...) While suspended telegraph wires are admittedly insulated differently, the way underground wires are insulated is comparable to that of nervous fibers" (Spencer 1883–1890: 82). In 1896, sociologist René Worms took up this metaphor of the nervous system-network in his book *Organisme et société* (1896), in which he likened "roads and railways to blood vessels, the telegraph to nerves, machines to the muscles of the social body" (quoted by Schlanger 1971: 90). This corporal fiction surrounding the network to make it an "organism-network" and export it into the social sphere was cyclically reformulated for electricity and telegraph networks, by Spencer, for the computer, by Neumann and Wiener's neuro-cybernetics, for telecommunications, by numerous engineers, for the Internet, by Joseph Licklider, and then by the ideologists of the Net, gathered around cult magazines like *Wired*. So much so, that in retrospect the contemporary celebrators of reticular techno-utopias noted that "there is an uncanny continuity in the wiring of the planet since the discovery and first application of electricity. The telegraph, the telephone, the Internet, the World Wide Web have followed upon each other as if they were stages in a single technological development", as Derrick de Kerckhove wrote (2000 [1998]: 197).

Engineers have relentlessly drawn on this imagery, particularly to present the developments of the telephone, IT and telecommunications. Marshall McLuhan was the one to link the prophets of electrical life with those of the digital revolution. He attributed a particularly important role to the electrical revolution which he saw as projecting our entire nervous system onto the world and bringing the world back to our nervous system. McLuhan wrote: "Today, after more than a century of electric technology, we have extended our central nervous system in a global embrace, abolishing both space and time" (McLuhan 2001: 21). And he added his famous sentence: "As electrically contracted, the globe is no more than a village" (ibid.). In *The Gutenberg Galaxy*, McLuhan borrowed Teilhard de Chardin's idea of "noosphere", which he interpreted as the "technological brain for the world, [...] the cosmic membrane that has been snapped round the globe by electric dilation of our various senses" (McLuhan 1962). Electricity inspired the author of *Understanding Media* to write pages of mystical inspiration on the electricity network: "In this electric age we see ourselves being translated more and more into the form of information, moving toward the technological extension of consciousness... By putting our physical bodies inside our extended nervous systems, by means of electric media, we set up a dynamic by which all previous technologies (...) will be translated into information systems" (McLuhan 2001). He summarized this fusion between electricity and the nervous system in a key sentence: "Electric technology is directly related to our central nervous system" (ibid.). The technical network and the biological network are extended until they are one, in accordance with the Cartesian marker of the reticular techno-utopia. These texts by McLuhan regarding electricity still fuel discourses on telecommunication networks and Internet.

The Communication Network: Society's "Nervous System"?

From the time it was born, in the 1880s, the telephone was seen as one of the "marvels of electricity" and extended the fictions fostered by electricity. Throughout the twentieth century the image of the "nervous system" promoted by Spencer was used by engineers and industrialists in telecommunications and IT to "give substance" to the network. This image was thus used by Theodor Vail, the head of American Telegraph and Telephone in its early days, for whom "the Bell system was developed in a spirit of intelligent control and as a large structure, to the point of merging with the nervous system of the country's economic activity and social organization (it even de facto became this nervous system)". This image of a society's nervous system is commonly used to define the telecommunications network, constituted of links and nodes, lines of transmission and commutators, all of which are increasingly "intelligent", to the point of being identified with the brain.

Reference to society's nervous system is recurrently used to define the network in the discourses of engineers and political figures on telecommunications. Some engineers, however, prefer to use the Saint-Simonian markers of the reticular techno-utopia – as was evidenced in a 1994 public report on "information highways" by Gérard Théry, former general manager of telecommunications in France – , and the Proudhonian marker to associate a type of social organization with a technical network architecture. The former approach emphasizes the socio-economic "revolution" brought about by the telecommunications network, to provide evidence of a political orientation, while the latter stresses the similarity between the reticular and organizational architectures, to promote new modes of management. The introduction to Gerard Théry's report starts with a sentence assembling the markers of engineers' reticular fiction, inherited from Saint-Simonism: "The revolution of the year 2000 will be that of information. While its technical scope will be comparable to that of railways or electrification, it will have more profound effects, for telecommunication networks now constitute the nervous system of our societies" (Théry 1994: 11). This statement establishes a causal link between the new technical network and social transformation seen as self-evident: telecommunication networks bring about a social revolution, for they are the nervous system of society. The use of Saint-Simonian markers of reticular fiction is facilitated by the fact that these networks are said to be "intelligent", for "information highways constitute the medium of the post-industrial society, essentially built on information exchange". This revolution "will fundamentally modify economic structures, modes of organization and production, each and everyone's access to knowledge, leisure, work methods and social relations". The Saint-Simonian marker of economic prosperity promised by the network is predominant in the demonstration. "Awareness of a new society emerging will thus develop"; this is a "post-crisis society, the information society" (ibid.: 121). for this techno-utopia conveys a vision of society to be achieved through the development of technical networks.

The famous 1978 Nora-Minc Report on the computerization of society, made the very direct assertion that: "Data processing offers the means to implement the most diverse schemes, the 'Tout-Etat' [complete centralization of state control] as well as that of extreme decentralization. Thus, guiding the acquisition of data processing means selecting a model of society" (Nora and Minc 1978: 105). In order to legitimate the 1982 "Cable Plan", the French government drew on telecommunications engineers' discourses to the letter and contrasted networks with "a so-called 'tree' structure, similar to those of water and electricity networks, able only to convey one-way traffic", with "radial networks, the structure of which is mapped onto that of the telephone distribution network, needed to evolve towards the offer of interactive services". Meanwhile, the evangelists of the Internet put the libertarian reference to the decentralized structure of the network, representative of a type of egalitarian society, at the heart of their discourse, which Christian Huitéma sums up as follows: "Unlike radio or television, the Internet is not a one-way medium. What is most revolutionary about the network is precisely the individual's capacity to be both a consumer and a source of information. (...) The Internet, far from being an institution of control, will on the contrary be an instrument of freedom, promising modern humans the ability to shake off the yoke of bureaucracies (...). In computerized businesses the emancipation of communication from hierarchical channels is already visible and, gradually, hierarchies are being flattened, fearful deference and arrogant certainty are giving way to egalitarian dialogue" (Huitéma 1996). The analogy between broadcasting networks and the hierarchical structure of a pyramidal organization leads to a critique of state centralization, which is likened to Orwellian controlled communication. The libertarian markers of reticular techno-utopia facilitate the transition from engineering to sociology, as networking structures inspire organizational or even societal models that seamlessly merge with one another. The two facets of the network, technical and organizational, are perfectly reversible thanks to the similarity of their architecture. The reticular form yields meaning, as a "graphical reason" applicable to diverse objects, some technical, others social.

Neuronal Networks and the Computer

The identification of the communication network with the nervous system was popularized by early cybernetics, which brought psychologists and mathematicians together, and particularly by Warren McCulloch (1892–1969) and Walter Pitts (1923–1969). In their famous 1943 article, "A logical calculus of the ideas immanent in nervous activity", they asserted that "the nervous system is a network of neurons" (1943: 62). As early as 1923, McCulloch had imagined an equivalence between the calculation of propositions and the rules underpinning the functioning of the nervous network. Meanwhile, in 1923–1924, Pierre Janet, professor at the Collège de France, wrote that "the brain is but a set of commutators" (quoted by Canguilhem 1993: 15). In their 1943 article, McCulloch and Pitts showed that "neuronal networks" are comprised of elementary interconnected "formal neuron" units: the neuron networks that constitute the cortex are formalized into "formal neuronal networks". They explained thought as the product of the brain's material structure, that is to say, of the neurons functioning as a network which allows for constant interaction within the brain. "A formal neuron", Henri Atlan commented, "is constituted of a body, or soma, from which outlets or axons lead to one or several endings divided into exciters and inhibitors. Each module of this kind receives endings from other modules..., via connections called synapses" (Atlan 1992: 129). The functioning of nervous activity may be formalized using propositional logic: "it seems that each network's behavior can be described in these terms if more complicated logical tools are added for networks containing loops; moreover for any logical expression that fulfils certain conditions, a network can be found that behaves according to the model described" (ibid.). The aim here is to constitute a logic through the propositional calculation of the "behavior of complicated networks". This is possible as "each neuronal reaction corresponds to an assertion of an elementary proposition" (ibid.: 64). McCulloch and Pitts' model associates a logical machine with a biological machine. Even if Mc Culloch and Pitts' formal neuronal networks are "schematizations of real neurons", resembling them somewhat, Atlan notes that they showed that "the functioning of the brain and that of artificial automata obey the same principles" (ibid.: 133). In fact, in the 1943 article, McCulloch and Pitts (1943) likened the brain to the computer: "the brain may be likened to a digital computing machine consisting of ten billion relays called neurons", therefore "the brain is a logical machine". However, since the human brain is "by far the most complex of data processors", the analysis of its mode of functioning will apply to any other complex system. Given that in the brain "each relay is a living cell", the referent of the two forms of "relay-cell" is a telecommunication network. Conversely, the brain remains a model for the engineer: even if "engineers cannot hope to compete with nature (...) computing machine designers would be happy to swap their best relays for nervous cells" (ibid.: 195).

As early as 1944, the computer was considered as an artificial brain, for the nervous system was the prevailing metaphor to think about electricity and telephone networks. Turing and Von Neumann dreamt of building a reduced model of the human brain, or at least an automaton whose functioning would obey a similar logic, an "electronic brain". Von Neumann saw a similarity between the functioning of a calculator's logic and that of the human brain: "the nervous system", he wrote, "is a computing machine which manages to do its exceedingly complicated work on a rather low level of precision" (1958: 78). This comparison between the brain and the computer is not self-evident, as it assumes an analogous functioning and architecture. Philippe Breton (1990: 140) highlighted two understandings which liken the computer to the brain: the one likens the comparable material infrastructures of the machine to those of the nervous system, as two material networks (the one machine, the other neuronal) designed to "provoke thought"; the other sees two comparable logical modes of functioning. In other words, does the neuronal network obey a binary logic, of the Boolean type, a condition of the logic programming cerebral and computer activity, or can it be a structure that produces intelligence in general? Turing and Neumann argued in favor of the latter. According to this hypothesis, which sees neurons as binary, the brain and the computer share the same logical functioning. This "computer scientist" thesis differs from the one supported by cyberneticists. Whereas Turing and computer scientists insisted on the logic of "intelligence", irrespective of its material medium, cyberneticists sought to construct artificial animals, by working on "the material medium of intelligence". In *Cybernetics and Society,* Norbert Wiener explained: "It is my thesis that the physical functioning of the living individual and the operation of some of the new communication machines are precisely parallel in the analogous attempts to control entropy through feedback" (1988 [1950]: 26). Comparable regulatory mechanisms exist in the organism and the machine: "This is the basis of at least part of the analogy between machines and living organisms. The synapse in the living organism corresponds to the switching device in the machine" (ibid.: 34). On this basis, Wiener compares the telecommunication network and the living organism: "there is no absolute distinction between the types of transmission which we can use for sending a telegram from country to country and the types of transmission which at least are theoretically possible from transmitting a living organism such as a human being", for "to be alive is to participate in a continuous stream of influences from the outer world" (ibid.: 141). Wiener's cybernetic human is situated at the heart of a network. Traversed by a network, he/she is plugged in, connected and communicative. "The representation of the human as a 'communicative being' is closely intertwined with the metaphor which associates the human brain and the computer", claimed Philippe Breton (1992: 52).

The Technology of the Reticular Mind

The technology of the mind, understood as a canonical reasoning process used in various disciplines, is the expression of the dispersion and commercialization of the concept that has become a "precept" with the fact of being in a network and thinking in terms of it. This deteriorated concept is a catch-all which, albeit useful in various disciplines, loses all substance by accounting for everything. Common to all its uses is the reduction of the network to the hidden structure of a system, a formalizable architecture made up of intertwined links or relations, in other words, interconnections. This structure tends to become the universal key to explaining the functioning of a complex system whatever it may be (society, brain, body, city, planet, world, etc.). The reverse is also true: detecting or imagining a network architecture in or under a complex system is enough to deduce its mode of functioning and transformation. The network defines a hidden order that can be acted upon. Mapping the uses of the word by discipline would reveal its presence in many disciplines: the information and communication sciences, the engineering and territorial development sciences, geography, but also history, the social sciences (network economics, management and the management sciences, the organizational sciences, sociology, political science), physics, biology, the mathematical theory of graphs, the cognitive sciences, etc. Across the board, the "hard" or "natural" sciences employ the network concept, which they seek to formalize using graph theory and "automata networks".[17] They apply these theoretical models and formal tools in various disciplines to explain complex systems.

The focus here is not so much on this formalization as on the relationships between the techno-utopia and the technology of the spirit of the network, in other words, on the interaction between engineers' discourse and the social sciences in addressing social functioning and change. The term "reticular expressivity" could be used to identify these intermediary narratives between the socialization of technical networks and the technicization of social change. These discourses draw on a hybrid technical-social definition of the network, play on its dual half-technical half-social character, and equate the technicization of the social to the socialization of techniques. The structure of the network thus plays a mediating role between technique and society. The network serves two purposes, as a technical matrix and as the structure of organization, even of the social realm as a whole. The network is at once a "technical network", an "organizational network", and the transition between the two. The technology of the reticular spirit at play in socio-economic discourses complements engineers' techno-utopias. It reveals the difficulty in conceptualizing the network other than metaphorically or by reducing it to a structure explaining a system.

The Pyramid and the Network in the Discourse on the Sociology of Organizations

In the late 1960s, Europe and the United States engaged in an intense trading of concepts to characterize the rise of the Fordist model of industrialism, and to outline its new "post-Fordist" type of organization, built around "a service economy". By 1967, the issue of "the information revolution" imported from the United States had already been widely popularized in Europe by Jean-Jacques Servan-Schreiber's *The American Challenge*, in which he wrote that "a technological revolution is underway. Its impact on modern society should be radical" (1967: 105–106). Meanwhile, US sociologist Daniel Bell, in *The Coming of Post-Industrial Society*, and a former adviser of the White House, Zbigniew Brzezinski, in *Two Ages: America's Role in*

---

[17] An *automaton* is a basic processor defined by three characteristics: an intense state, connections (with other automata or an environment) and a transition function allowing it to calculate its internal state based on the signals it receives about its connections.

the Technetronic Era, both theorized this transition towards "post-industrialism".[18] Daniel Bell saw the concept of post-industrial society as the outcome of a change in the social structure through technology: "The aim of formulating the concept of post-industrial society is to highlight a change in the social structure"; "the notion of post-industrial essentially refers to the transformations of the social structure" (ibid.: 153 and 418). Yet, "insofar as social evolution is linked to that of technology, the major changes in the next fifty years will stem from the telecommunications revolution. From 1825 to 1875, we experienced half a century of British supremacy: it was the fruit of the railway revolution" (ibid.: 428). Bell argued that technical transport or telecommunications networks transform society through its structure. This idea, directly inspired by the Saint-Simonian approach, very quickly gained currency. It was developed as "hyper-industrialism" by Alvin Toffler and John Naisbitt, as "post-modernity" by Lyotard, and as "network society" by Manuel Castells who updated the approach in the era of Internet. However, many authors mediated between the Saint-Simonian discourse on industrialism and Bell's new take on post-industrialism. According to Bell himself, these authors included US sociologist Thorstein Veblen with his 1919 book *The Engineers and the Price System*, Wiener with cybernetics, and James Burnham, who played a key role in the sociology of organizations, with *The Managerial Revolution: What's Happening in the World*. The definition of contemporary society as a transition towards "post-industrial" society was transferred into managerial discourse in the 1970s by Alvin Toffler who, with hints of Neo-Saint-Simonianism, researched "the transition from industrialism to super-industrialism" (Toffler 1985). Toffler explained this transition as follows: "Industrialist bureaucracies have a pyramidal structure (...)", whereas "any country that turns the page of the industrial chimney era needs decentralized, ultrafast networks with a high capacity to circulate considerable masses of computer data, video images and other types of messages, alongside conventional telephone calls" (Toffler 1986: 135 and 143). The Nora-Minc report followed suit, transferring into the political field this theme of the "shift from the organic industrial society to the polymorphous information society" (Nora and Minc 1978: 114), brought about by the "CTI revolution". In this report, the network is considered in ambivalent terms, both as a technique for the circulation of information and as a mode of social organization. It serves two purposes, as an information technique and an organization structure: "The challenge (...) lies in the difficulty of building the system of connections that will allow information and social organization to progress together" (ibid.: 16). The main assertion is that the network changes organizations and society as a whole. In the same year, in *The Network-Nation: Human Communication via Computer,* Starr Roxanne Hiltz and Murray Turoff (1978) discussed "the network nation" which could "re-unite individuals and groups dispersed over wide distances... and recreate emotional bonds";

---

[18]"Daniel Bell was unchallenged as he launched the concept of 'post-industrial society'. This notion was already at least implicit in the book he published in 1960, entitled *The End of Ideology*", wrote François Bourricaud in the preface from the French edition (Bell 1976).

thanks to group communications, "we will become the Network Nation, exchanging vast amounts of (...) information". The authors argue that images of communication networks and social organizations can be reversed, as the network corresponds to a dual technical-social structure. This allows engineers to socialize techniques, and sociological discourse to make the social technical. The Proudhonian network marker acts both ways: a network architecture reveals a choice of social organization and, conversely, the social organization becomes adequate for the technical networks it uses. Two structures are systematically contrasted as symbolic figures of power by engineers and sociologists specialized in organizations alike: the vertical pyramid and the horizontal network. In the early 1980s John Naisbitt pointed out in *Megatrends* that one of the "ten new directions transforming our lives" is the transition from "hierarchies to networking: For centuries, the pyramid structure was the way we organized and managed ourselves"... "From the Roman army to the Catholic Church, to the organization chart of the General Motors and IBM, power and communication have flowed in an orderly manner from the pyramid's top, down to its base (...) The reticular model we have now all adopted with extraordinary success is replacing the hierarchical form" (Naisbitt 1984: 247 and 251). Naisbitt advocated "destroying the pyramids" through "networking", for the network, he claimed, would ensure what "the bureaucracies can never provide: horizontal law" (ibid.: 247 and 255). This discourse on reticulated organization, popularized by management and futurology discourse, found theoretical benediction in *The Postmodern Condition* by Jean-François Lyotard, who explicitly followed in the footsteps of Touraine and Bell with their "post-industrial society", and in the context of the frequently referenced Nora-Minc Report. Lyotard argued that the precondition of post-modernity is the fragmentation and disaggregation of the social and its dispersion into "clouds of sociality" which reconstitute themselves into "intersections" where "each of us lives" (1979: 8). The figure of the network is presented as a flexible rearrangement of the social pyramid, following its prior disintegration into clouds. This reticulated figure founds (and is founded upon) a new technological legitimacy, that of informational and computer-telephone integration (CTI) networks: "Where, after the metanarratives, can legitimacy reside? The operative criterion is technological" (ibid.). Consequently, wrote Lyotard, "no self is an island; each exists in a fabric of relations that is now more complex than ever before... located at 'nodal points' of specific communication circuits" (ibid.: 31). In short, the networked technological prosthetics is the product of social disintegration. Commenting on Lyotard's contribution, philosopher Dominique Lecourt rightfully highlighted the importance of the technicist discourse on networks in Lyotard's work: "The postmodern mind readily worships technology... The lesson it offers is clear: the new information and communication techniques provide a powerful contribution to networking a society that has become decentered, de-pyramidalized, tormented with countless unstable flows allowing individuals' activities to unfold over the course of a more or less exhilarating nomadism" (Lecourt 1999: 106–107). The "post-industrialist" and "post-modernist" discourses are extended to saturation point by the management and economic discourse on "post-Fordism", which celebrates enterprise and "decentralized" networked organizations. The hierarchical industrial enterprise gives way to the flexible, relational, contractual, networked enterprise. The neo-Fordist enterprise brands itself "network-company": organized with, by and into networks. It contrasts with the "pyramidal" Fordist enterprise organized like a "castle". Federico Butera thus wrote: "we are currently leaving the model of the "castle" scientifically described by Max Weber, developed and implemented by Mary Theresa of Austria and Henry Ford, perfected in detail by Taylor and Fayol... The new organizational model is that of the network" (Butera 1991: 14). All the markers of the reticular techno-utopia are mobilized to legitimate this "reticular management", starting with the corporal metaphor, as illustrated by Georges-Yves Kerven who claims to have philosophically founded a discourse on the network company: "The company may also be analyzed as a network connecting brains together. The company thus appears like a network of brains, themselves networks of neurons. The company is therefore a network of networks... and resemblances exist between the brain of neurons and the network of brains, insofar as the brain analyses and structures the company as a network of brains" (Kerven 1993: 138). In *Face à la complexité, mettez du réseau dans vos pyramides* (*Faced with complexity, put some networking in your pyramids*) (Sérieyx et al. 1996), in which he popularized these images of the structure of organizations, Hervé Sérieyx meted out one of his managerial sentences: "The network is becoming the favorite mode of action of the era of intelligence, of complexity". He summarized this managerial ideology in a few slogans which are revealing of contemporary discourses on the network: "the pyramid divides up the work and at best adds up the tasks; the network multiplies the added value of contributions. The pyramid is frozen; the network benefits from a variable geometry. The pyramid centers itself on its own functioning; the network forever coevolves with its environment"; "The pyramid was the tool of manufacturing, the network is the tool of brain-factoring" (ibid.: 13 and 15). The pyramid thus relates to a "mechanic model" and the network to a "biological model" (ibid.: 14). The network is likened to the organism and contrasted with the pyramid-machine: "The network organization is the complete opposite of the pyramid organization: its development is cellular, the cell adapts, grows and divides to survive by transmitting its genetic code, just as living systems do", Sérieyx explained (ibid.: respectively 95 and 15).

Just as the engineer uses the organic metaphor to naturalize the technical network, so too the economist uses it to naturalize the market. This model of reticular management ideology was critiqued by Luc Boltanski and Eve Chiapello, who saw the 1990s as the time of triumph of the "model of the firm as network", because "hierarchy is a form of co-ordination to be excluded" (1999: see 111–123). This "rejection of hierarchy" is thought to afford autonomy and formal equality. The network-company is thought of as a fabric of interconnected autonomous projects, and the manager becomes the symbolic figure of the "networker". The network is presented as the technology of the spirit which allows for the encounter of two "post" ideologies (industrial, modern, Fordist): communication and management. As it links technical communication networks with organizational management networks, I propose a neologism, "comm-management" (Musso 2000) to refer to the intermingling of these two ideologies.

The Actual Components of Retiology

Retiology, the contemporary ideology of the network, merges the techno-utopia and the technology of the reticular spirit. It combines a deteriorated and worn-out concept and symbolic operation, to celebrate the new technical networks and convey the promise of transformations in society, customs, services, organizations, the economy, territories, etc. *Retiology* produces an inflation of intertwined images and discourses. This imagery surrounding reticular techniques and technologies supports the industrial propaganda of the *efficiency* (Legendre 2001: 59) and the "visionary" discourses on the future of the network society. *Retiology* is an ideology with utopian aspirations, a technological utopia, in other words a utopia whose referent is reduced to the fetishism of technical networks, particularly Internet and teleinformatics networks. *Technolâtrie* (worship of technology), "techno-*imaginary*", "techno-messianism", "techno-utopia": all these terms refer to this fetishism of the technical network that is meant to illustrate a "hidden God", creator of new social links, new communities, or even a new society. As Georges Balandier aptly put it, "The very modern image of a networked world can thus conjure up again other, very ancient images, through which lost or exotic civilizations have defined or still define their world as a complex and therefore fragile and uncompleted fabric" (2001: 14). *Retiology* takes as its object what Balandier calls the "encapsulated social, in other words caught in the envelope of global networks" (ibid.: 37). Its interpretation of the "social fabric" and its fate refers to its ultra-modern technical weavings, the Web, the World Wide Web. *Retiology* is the contemporary ideology of the Web, it relates not only to Internet, "the network of networks", but to all works whose elements intertwine and interconnect.

Drawing on the Saint-Simonian markers of reticular fiction, *retiology* announces the future society – the "post" society – already at play in the construction of technical networks, the *imaginaries* and the practices they bring about. *Retiology* constitutes a set of discourses and imagery, of "theorized practices" of networks, if not to say the claim of constituting a discipline. It already has "retiologists" and takes on the task of managing this transition and tension towards a promised future which unfolds in various ways: sometimes through the generalized liquefaction of the social, for example in the cyberspace woven by Internet and the social networks that see to the creation of communities; and sometimes through universalized fragmentation, then global reweaving, as for example in Manuel Castells' "network society". Cyberspace and "network society" are both figures that were constructed as reflections of the Internet, and constitute the two dimensions of *retiology*, that is, two enactments of the "social fabric" with the help of the Web. Whether in the form of cyberspatial literary fiction or socio-economic analysis of the network society, *retiology* forever announces "revolutions of (and through) networks". To this end, *retiology* draws on ancient imagery of the reticular narrative which facilitates the projection of the network society and cyberspace into the future. As a utopian ideology, it produces and reproduces old futures. The fetishized figure of the network, the object of the "new techno-worship" (expression from Balandier 2001: 34), always alludes to the "passage", in two main forms: the transition towards another state, and constant movement. The network sometimes refers to a future society rewoven by networks, in which case it serves as a sort of horizontal cathedral of post-modern times, and at other times it is just the sign of permanent transformation, of movement per se, in which the present society is constantly caught.

So as to question contemporary *retiology*, let us examine its two faces of Janus: on the one hand, its literary cyberspatial variant, advocated by cyberculture, and on the other, its sociological variant that defines the transition towards a "network society" and "informational capitalism", defended by Manuel Castells.

Cyberspace or Generalized Liquefaction

The techno-utopia of cyberspace, the Internet's contemporary twin, sounds the triumph of the Galenic-Cartesian marker of the network. Cyberspace conveys the image of a universal network connecting all the individual brains plugged in on a global scale which, according to "retiologists", constitute a sort of "global brain" as Joël de Rosnay (2000) calls it, which produces a "collective intelligence", to use Pierre Lévy's term (1999, 2001). In fact, this techno-utopia was built by Joseph Licklider, a psycho-sociologist working with MIT engineers, in a 1960 article called "Man-Computer Symbiosis". Licklider took John von Neumann's work on cybernetics in another direction, with the dream not so much of creating a machine that would be the brain's duplicate, but of interconnecting the brain and the computerized machine: "The hope is that, in not too many years, human brains and computing machines will be coupled together very tightly" (Licklider 1960: 4). He sought a "partnership" rather than a substitution between the brain and the computer: what he called a "symbiotic relation between a man and a (...) machine" (ibid.: 6). That is why he envisaged the creation of an IT network for generalized exchange between humans and computers. In 1968, Licklider co-authored an article with Robert Taylor, head of the ARPA's IT center, from which the Internet was to emerge. In this article they predicted that "men will be able to communicate more effectively through a machine than face-to-face... life will be happier for the online individual because the people with whom one interacts most strongly will be selected... communication will be more effective and productive, and therefore more enjoyable" (quoted by Flichy 2002: 41). This techno-utopia makes cyberspace a place where brains and computers are plugged into one another. To this end, they are both broken down into identifiable parts (the electronic chips equivalent to neurons) and "interconnected" to produce a small "intelligent" totality (the brain and the computer) that can be extended into a "large totality" built by analogy, that is, the "global brain" (linking up the interconnected brains and computers), endowed with "collective intelligence". This series of metaphors leads to a twofold identification: the brain is a computer and, like the computer, the brain has a reticular neuronal structure which supports intellectual activity.

The founding syllogism of the cyberspatial techno-utopia boils down to the following assertions:

1. the brain functions like a computer and, conversely, the computer functions (and "thinks") like a brain. Both ideas stem from a unitary theory concerned with the connection of networked elements: the brain is a network of neurons and the computer is comprised of networked chips;
2. with the Internet, a global network of networks is developing through the connection of the computers comprising it;
3. as a result, it is possible to link up human brains and computers through hypernetworks connected on a global scale. This affords the possibility of human-machine hybridization and "collective intelligence" in and through cyberspace. The construction of cyberspace relies on three assumptions: the network understood as a generalized interconnection, the existence of isolatable elements that are both different and similar, that is to say, brains and computers, waiting to be networked together, and lastly the human-machine hybridization, due to the brain-network-computer equation. In fact, the connectionist models legitimate this analogy. As Bechtel and Abrahamsen emphasized, "the initial impetus for developing network models of cognitive performance was the recognition that the brain is a network" (2002). Once these preconditions have been set, implicitly based on the markers of the reticular techno-utopia, cyberspace produces all the "beneficial" effects that "retiologists" are forever promising us. The main virtue of cyberspace is that it dissolves all disturbing elements – territory, institutions, particularly the State, and the physical body – and favors a quasi-religious asceticism regarding spirituality, enabled by the technical network of the Internet. Kevin Kelly, former editor-in-chief of *Wired*, the cult magazine for Internet users, thus described his first visit on the Internet as a "religious experience" (quoted by Dery 1996: 47). In 1992, John Barlow, founder of the *Electronic Frontier Foundation*, wrote that "The idea of connecting every mind to every other mind in full-duplex broadband is one which, for a hippie mystic like me, has clear theological implications" (quoted by Flichy 2002: 111). Cyberspace rearticulates the religious order and inserts it into technologies. In a sense it is the limit of reticular thinking, in its spiritual version. The establishment of cyberspace as an unlimited space for informational networks affords unrestricted movement in a pure space that is free of friction, ethereal and virtual. By way of exorcism, everything becomes possible in this ideational-ideal space, once territory has been forgotten. Jeremy Rifkin thus asserted that "The shift (...) from geography to cyberspace represents one of the great changes in human organization", even referring to "migration of territory to cyberspace" (Rifkin 2000: 17), for in cyberspace borders disappear, as does physical territory... The physical body also becomes superfluous, as only the brain is

engaged in the cyberspatial adventure. In the fiction of William Gibson, who in 1983 created the term "cyberspace" in his founding novel *Neuromancer* (Gibson 2011 [1983]) it is all about "neuroconnection". His definition of cyberspace is built on reticular imagery and... connectionist network architectures: it is *a* "consensual hallucination experienced daily by billions of legitimate operators, in every nation (...). A graphic representation of data abstracted from the banks of every computer in the human system. Unthinkable complexity". The hero Case, a hacker on the run, connects with cyberspace through a neurological interface, by plugging his nervous system into "the matrix", a global virtual reality where information is stored in the form of tangible illusions. Case lived "for the bodiless exultation of cyberspace", "jacked into a custom cyberspace deck that projected his disembodied consciousness into the consensual hallucination that was the matrix". In other words, Case experienced disembodiment – "the body was meat" – and was able to leave his body to journey in the cyberspatial yonder, guided by the ghost of a dead computer hacker, synthesized by a computer. Gibson imagined Case's brain and nervous system connected to the electronic network, cyberspace: the brain is externalized (into a computer – artificial brain), then connected. This interworld between technique and the body, between brain and network, is where the theme of the "wirehead" emerged. In *Schismatrix*, Bruce Sterling called the mechanists with prostheses who were connected via a computer "wireheads", and in cyberpunk circles the term was synonymous with "aspiring cyborg", since the cyborg is the connection of the individual brain to the global artificial brain (Dery 1996: 354, note 179). Meanwhile, the editor-in-chief of the journal *Mondo 2000* declared: "I think we're going through a process of information linkup toward the building of a global nervous system, a global brain" (quoted by Dery 1996: 47).

In the cyberspatial interworld, the technicized bodies and naturalized techniques are merged into a single term and into hybrid beings that resemble technical fictions. What makes the unity of cyberspace, if not the idea of "interconnection" with reference to communication networks, encapsulated in Joël de Rosnay's definition of the term: "cyberspace [is an] electronic space-time created by the emergence of communications networks and multimedia computer interconnections"? (de Rosnay 2000: 283). Cyberspace is a space of mechanical and organic networks interlinked ad infinitum, without borders. Pierre Lévy confirmed this reduction of cyberspace to the "network", then to the vague idea of interconnection: "Cyberspace (also known as the 'network') is the new medium of communications that arose through the global interconnection of computers... One of the ideas, or rather one of the strongest forces behind the development of cyberspace, is that of interconnectivity... Interconnectivity weaves a universal through contact" (Lévy 2001: xvi and 107–108). For the *retiologist*, interconnection ultimately amounts to the intuition of a "sensation of all-encompassing space". This sensation is strangely reminiscent of the "communion", as understood in its etymological and religious sense, as sharing or pooling. In cyberspace, rough and resistant territory is erased; only a smooth, fluid space remains that is made for circulation, a space of informational networks and links, without memory or places. This "space of space" of extended networks is hybrid, half-human half-machine. It indiscriminately connects humans and machines, as networks are sometimes brains and sometimes artefacts. Reticular fiction thus merges the technical and the biological into a seamless whole. Cyberspace is a hybrid being, but one that is "alive", as *retiologists* assure us. In this respect, postmodern philosopher Manuel de Landa wrote: "Past a certain threshold of connectivity, the membrane which computer networks are creating over the surface of the planet begins to 'come to life'. Independent software [programs] will soon begin to constitute even more complex computational societies in which [programs] trade with one another, bid and compete for resources, seed and spawn processes spontaneously, and so on" (quoted by Dery 1996: 44). Meanwhile Pierre Lévy declared that "cyberspace is similar to certain ecological systems"; "Its center will be everywhere, and its circumference nowhere. This hypertext computer will be dispersed, living, pullulating, incomplete: cyberspace itself" (2001: 93 and 26). Likewise, Joël de Rosnay's cybion is "A hybrid biological, mechanical, and electronic super-organism that includes humans, machines, networks, and societies" (2000: 132).

Cyberspace is a powerful symbolic dissolver – a "consensual hallucination" –, as it eliminates all sources of resistance: the territory, the body, but also politics and the State. Thanks to the network, democracy will be electronic and "the political will disappear", Jacques Attali announced.[19] Through the generalized liquefaction brought about by cyberspace, the political and its state-national form can be eliminated. Manuel Castells declared that "networks destroy state control over society and the economy. What is over, at this current stage, is the Sovereign, national State".[20] As early as 1979, Jean-François Lyotard announced that "The ideology of communicational 'transparency', which goes hand in hand with the commercialization of knowledge, will begin to perceive the State as a factor of opacity and 'noise'" (1979: 15–16). This liberal-libertarian anti-state vision, inherent to web surfers' ideology, merely updates the Proudhonian marker of the reticular techno-utopia. The network, considered to be anti-hierarchical "in essence", becomes synonymous with self-regulation and equality. That is why the *Internaut* (web surfer) is meant to fight for freedom against all regulatory organs, against the dominant players (Microsoft, Google or the FBI, for example), for equality against all hierarchies, starting with those of States, and for the global fraternity of "virtual communities". Freedom, equality and fraternity: the social utopia of 89 (1789–1989) is said to finally be here, thanks to the technical reticular utopia. As Pierre Lévy put it: "Cyberspace appears as a kind of technical materialization of modern ideals" (2001: 230). Certain evangelists of the "New Age" have found the same virtues in the network. Marilyn Ferguson writes that the network is "the antidote to alienation. It generates power enough to remake society. It offers

---

[19]Jacques Attali, *Libération,* 12 June 1998.

[20]Conversation with Jacques Attali in the "Multimédia" supplement of the newspaper *Libération,* 12 June 1998.

the individual emotional, intellectual, spiritual, and economic support. It is an invisible home, a powerful means of altering the course of institutions, especially government. The Aquarian Conspiracy is (...) a network of many networks aimed at social transformation" (Ferguson 1987: 213).

Derrick de Kerckhove also celebrated the "connected intelligence" and saw "the essence of any network" (2000: 18) in webitude or the "mental bond between people", for the Internet "gives us access to a live, quasi-organic environment of millions of human intelligences". Kerckhove explicitly supports a biotechnological vision of the network, merging technical and biological networks: "continuity between the two domains, the technological and the biological, is established by the fact that there is electricity both inside and outside the body" (ibid.: 196). The author reveals the value of the organic model for *retiology*: the concern is to provide the unity, regulation and social totality of integration that gives substance to technology. "With the appearance of the Internet integrated on the scene, it is as though technology discovered a way of imitating the physical, biological body in the social, technological domain: each party is connected to all the others to ensure the integrated functioning of the whole" (ibid.: 200). In particular, Derrick de Kerckhove provided a key to decipher *retiology* when he wrote that "One of the main effects of digitization is to make 'liquid' everything that is solid" (ibid.: 196). Digitization into bits of information has allowed *retiology* to atomize the real and transform it into a fluid that circulates within networks. According to Kerckhove, the ultimate stage of this liquefaction is the transmutation of these bits into thought: "This very flexibility makes matter, once perceived as consisting of mutually heterogeneous and impenetrable substances, seem now as fluid as thought itself (...) The spirits on the Net are connected and do behave like liquid crystal in stable though fluid formations" (ibid.: 205). Beyond the "digital man" so dear to Nicolas Negroponte, the "digitization of bodies" is at play. With the cumbersome and imperfect body liquefied or reduced to a digital bank, comes forth "Homo silicium", to use David Le Breton's expression (1999: 201). In fact, cyber-liquefaction leads to liquidation of the body, purely and simply, that is, according to Yves Stourdzé, to "corporal extermination" (1998: 142). But internet *retiology* can be pushed to the point of technico-spiritualist delirium: Jean Houston, a philosopher and historian of culture who co-runs the *Foundation for Mind Research* in New York, claimed that "if the Internet is a product of divine creativity, even as we humans are, perhaps in some sense, it is a new life form, a silicon-based living being which may be one of our evolutionary descendants. And yet, the very biology of its biosystem is mystical in nature – a vast, nonlinear reality wherein, like Indra's Net, each node connects to every other. Its webbed world encompasses the accouterments traditionally assigned to the Mind of the Maker – circles, nets, infinite feedback loops, the endless flow of being and becoming, God's identity as that perfect sphere whose center is everywhere and whose circumference is nowhere. Add to this the Net's ever-unfolding pattern of novelty, and we have a living system, one which reflects the nature of life in all its iterations" (Houston 2000). Although *retiology* reaches its extreme form here, with techno-devotion as the mystical delirium of the network, it can also take on more rational forms, still however relying on the fetishism of the

Internet, to herald a social revolution. While the Internet fluidifies the social and bodies, through generalized digitization in cyberspace, it also recomposes links in a fragmented society that it networks, according to Manuel Castells. Digitization and fragmentation are the preconditions for intervention by the reticular prosthetics that reweave "spiritual" links in cyberspace, and "material" ones in the "network society". Castells argues that it is not so much a matter of fluidifying society and the territory – as in cyberculture, which nevertheless remains a reference in his demonstration –, as thinking about social change, announcing the transition between a society in crisis, under "financial capitalism", and a new society, under the networked "informational capitalism".

Manuel Castells' Network Society

In *The Rise of the Network Society*, the first volume of his trilogy titled *The Information Age*, Castells presented a vast synthesis of the techno-utopia and the reticular technology of the spirit in the Internet age. He thus provided a comprehensive retiological survey. Starting with the "Internet revolution", he drew on the full range of markers of the reticular techno-utopia and used an elastic notion of the network that took on no less than twenty meanings before completely emptying itself out in a final definition of "interconnection", shared with cyberculture. Yet the notion of "network" is crucial to his entire demonstration, which is based on the axiom of "the pre-eminence of social morphology over social action" (Castells 2010 [1996]: 500). The notion of network – of which the Internet is a "pure" example – is presented as the determining structure of society: "The convergence of social evolution and information technologies has created a new material basis for the performance of activities throughout the social structure. This material basis, built in networks, earmarks dominant social processes, thus shaping social structure itself" (ibid.: 502). If the network was removed (like "pulling the rug from under his feet"), his argument would collapse. Castells' articulation of the network stems from retiological belief and epistemological fuzziness. The author begins with the following statement: "A technological revolution, centered around information technologies, began to reshape, at accelerated pace, the material basis of society" (ibid.: 1). Castells is concerned with the Internet and interconnected computer networks: "Interactive computer networks are growing exponentially, creating new forms and channels of communication, shaping life and being shaped by life at the same time. Social changes are as dramatic as the technological and economic processes of transformation" (ibid.: 2). The paradigm of the network is obviously the Internet: "The Internet is the backbone of global computer-mediated communication". It is even THE archetypical network, for it is the "network of networks" (ibid.: 375 and 383 respectively). This McLuhanian or even Neo-Marxist statement – the technical revolution affects society through its material structures – is but a repeat of the first Saint-Simonian marker of the techno-utopia, which is that the network heralds a technical and social revolution. The Internet network ensures the shift from technical change to social transformation. Castells actually later dedicated a book to "the Internet Galaxy", with an explicitly McLuhanian title, in which he argued that "The Internet is the technological basis for the organization form of the Information Age: the network" (Castells 2001). He drew on this fetishism of reticular technique as an argument: "The story of the creation and development of the Internet is one of an extraordinary human adventure", he wrote in *The Internet Galaxy*, in which he frequently used the adjective "extraordinary" to describe the Internet which, he added, "is indeed a technology of freedom" (2001: 1, 9 and 275 respectively). This approach, which affirms the existence of a base and technical infrastructure supporting the whole social fabric, is driven by a mechanistic vision: "The Internet provides the material basis for these movements to engage in the production of a new society" (ibid.: 143). Castells aptly summarizes the scope he attributes to the Internet, that is to say, the generalized networking of society, power and organizations: "The Internet (...) is not just a technology. It is the technological tool and organization form that distributes information power, knowledge generation, and networking capacity in all realms of activity" (ibid. 269). The Internet captures the whole social realm in its nets, the Web redefines the social fabric, as the railway or electricity once did. The Internet network is both the invisible social link (its hidden material structure) and the subject of the digital "revolution". The author refrains from any technical determinism, though he does state that his starting point is "the process of revolutionary technological change" (Castells 2010 [1996]: 4) and that "technology does not determine society. Nor does society script the course of technological change" (ibid.: 5). He observes a complex set of interactions. Despite this denial, Castells' reasoning is still underpinned by technological determinism: the technological revolution is that of IT networks. However, since the material basis of society is comprised of technological networks, society enters a revolution that constitutes the "general overhauling of the capitalist system" (ibid.: 2). Articulating all these Saint-Simonian markers of the reticular techno-utopia, Castells heralded a plethora of changes, as the mechanical consequences of the "effects" of the network defined as the material and cultural structure of the "informational capitalism" that he saw emerging. Castells' "informational capitalism" pursues the ideas of Alain Touraine (who prefaced the translation of Castells into French) and Daniel Bell on post-industrial society, and re-examines the idea of the information society, in the Internet era. He describes this "informational capitalism" as the combination of a mode of production – financial capitalism – and a mode of development linked to the Internet. Based on this technical-economic paradigm, he sees a "new society emerging from this process of change [that] is both capitalist and informational" (ibid.: 13). Wary of veering into futurology, Castells nevertheless uses the technical network's capacity to present itself as a transition towards an information society to come: the network society "emerging as a transitional form toward the informational mode of development that is likely to characterize the coming decades" (ibid.: 78). He characterizes the information society in terms of social fragmentation, the "general destructuring of organizations", and the isolation of individualities; the social link could (and should?) consequently be reconstructed using technical reticular prosthetics. Castells posits the prior atomization of the social before heralding its salutary "networking". The new weaving of the social fabric is operated by the Internet: "The novelty is networking through the Internet" (2001: 176). The demonstration is based on the constitutive social atomization/technical network duo. The image of the network is the reverse of that of the demise and fragmentation of society: "we observe (...) throughout the world, (...) the increasing distance between globalization and identity, between the Net and the self", Castells wrote (2010 [1996]: 22). To support his assertion of the network's superiority over social atomization, Castells simply cites Kevin Kelly (ibid.: 70),[21] one of the popes of cyberculture, who stated: "The Atom is the past. The symbol of science for the next century is the dynamic Net... Whereas the Atom represents clean simplicity, the Net channels the messy power of complexity... The only organization capable of non prejudiced growth, or unguided learning is a network. (...) Indeed, the network is the least structured organization that can be said to have any structure at all... In fact, a plurality of truly divergent components can only remain coherent in a network. No other arrangement – chain, pyramid, tree, circle, hub – can contain true diversity working as a whole". Castells' socio-economic demonstration also draws arguments from cyberculture and its fictions, as the two share a belief in *retiology*. Castells explicitly supports cyberculture as a suitable culture for the organization of the network enterprise, the cornerstone of this new capitalism: "there is indeed a common cultural code in the diverse workings of the network enterprise. (...) It is a *multi-faceted, virtual culture*, as in the visual experiences created by computers in cyberspace by rearranging reality. It is not a fantasy, it is a material force" (2010 [1996]: 214).

Since the notion of network is the cornerstone of Castells' reasoning, one might expect a rigorous definition. Yet it is limited to the following, provided in the conclusion of the book *The Rise of the Network Society*: "A network is a set of interconnected nodes. A node is the point at which a curve intersects itself. What a node is, concretely speaking, depends on the kind of concrete networks of which we speak" (ibid.: 501). The same definition is used again on the cover of *The Internet Galaxy*: "A network is a set of interconnected nodes" (Castells 2001: 9). This minimalist definition of the network, reduced to a function of interconnection, is so weak as to be applicable to any object whatsoever. Only the connection remains. Castells thus multiplies the uses of the word network, which takes on no less than twenty different meanings in *The Rise of the Network Society*, securing the unity of the analysis through shifts in meaning. Just as Diderot's *Encyclopaedia* observed "network effects" everywhere in nature, so Castells notes the generalized networking of the social, thanks to reticular techniques.

There is no better way of illustrating the deterioration of a concept into a technology of the spirit than through this enumeration intended to support the techno-utopia of the Internet revolution, which has become "the lever for the transition to a new form of society – the network society – and with it to a new

---

[21] Manuel Castells' citation is in a note (2010 [1996]: 70, note 87). See Kelly (1995).

economy" (ibid.: 2). According to the Saint-Simonian antiphony, the new technical network is fetishized as bringing social change: "Presence or absence in the network and the dynamics of each network *vis-à-vis* others are critical sources of domination and change in (...) the network society" (2010 [1996]: 500).

Castells' generalized networking of the social in response to its prior atomization echoes the generalized fluidification of the social – "the liquid society" – imagined by cyberculture, thanks to numeric digitization. The present society, scattered and fragmented, can be regenerated thanks to the network, either through generalized fluidification, or through reconstruction in a new social fabric. The reticular techno-utopia is always transformative, but in two different modes that define the network's "double body": the passage-transition from one state to another, or the continual passage-flow and movement.

Conclusion

Retiology is prevailing as a contemporary ideology, thanks to technical determinism. George Balandier was right to note that "everything seems to converge towards the most complete realization of the Saint-Simonian prophecy: to replace the government of people with the administration of chattel and the Organization" (2001: 254). However, this realization of Saint-Simonian *New Christianity*[22] is far more of an administration of chattel turned government of people.

Drawing on the markers of reticular fiction, *retiology* is forever heralding the future (or "post") society already at play in the construction of technical networks and the *imaginaire* they convey. It constitutes a set of discourses and imagery, or "theorized practices" of networks and even claims to constitute a discipline. Moreover, it already has its *"retiologists"* and has taken on the task of defining this transition towards the promised future, which is said to follow two main paths: either through the fluidification and generalized digitization of the social whole, for example in cyberspace, or through global reweaving, for example in Castells' "network society". These two facets of *retiology* suggest the restoration of the social link through the binding and regenerating virtues of technical networks. The world will either be fluid and liquid (Zygmunt Bauman), or a "feudalization of networks" (Pierre Legendre).

Contemporary *retiology* recycles and carries into the future an old imagery of the reticulated, burdened with a long history. It produces and reproduces old futures. The fetishized figure of the network, the object of its "new techno-devotion" (Balandier 2001: 34) always relates to a shift, in two main ways: the transition towards another state to come, or immediate motion. The network alternately refers to a future society rewoven by networks, or to movement per se, within which

---

[22]Title of Henri Saint-Simon's last book, 1825. See *Œuvres complètes* by Henri Saint-Simon (Grange et al. 2012, vol. 4).

individuals and society are constantly steeped. *Retiology* thus articulates two forms of transition understood as a crossing towards a new state or in immediate immersion in flows. Jean Baudrillard thus observed that "We are networked, we are the network. (...) We are steeped in it. Our present merges in with the flows of images and signs. (...) We are in real time".[23] The movement is continuous. There is no longer any need to bring about social change; it is constantly experienced through the connection or "plugging into" the networks. This "post-modern" staging of transition is thus experienced in the practices and rites of places of transition and communication, which Marc Augé called "non-places": doors and access keys, security doors, security gates or connection gates, to manage the daily ceremonies of entrances-exits in networks.

To enchant the generalized embrace of bodies, cities, society and the entire planet by technical energy, transport and communication networks, contemporary *retiology* celebrates the achievement of techno-utopia in the daily practices of circulation in networks and of connection to networks. It thus interlinks discourses and images of the reticulated to account for the contemporary "social fabric" and legitimate industrial propaganda in favor of the development of technical networks.

Retiology is an ideology with utopian aspirations, which is limited to the fetishism of technical networks, particularly the Internet. Whether it be literary fiction, futurology or socio-economic analysis of the "network society", *retiology* is constantly heralding socio-technical "revolutions". It thereby relieves social and political utopias of their heavy burden by transferring it to the technological utopia, which has the advantage of always materializing.

References

Atlan, H. 1992. *L'organisation biologique et la théorie de l'information*. Paris: Hermann.
Balandier, G. 2001. *Le grand système*. Paris: Fayard.
Bechtel, W., and A. Abrahamsen. 2002. *Connectionism and the mind: Parallel processing, dynamics, and evolution in networks*. New York: Wiley.
Bell, D. 1976. *Vers la société post-industrielle: essai de prospective sociologique*. Paris: Editions Robert Laffont.
Bertin, J. 2010. *Semiology of graphics: Diagrams, networks, maps*. Redlands: ESRI Press.
Boltanski, L., and E. Chiapello. 1999. *Le nouvel esprit du capitalisme*. Paris: Gallimard.
Bressand, A., and C. Distler. 1986. *Le prochain monde: Réseaupolis*. Paris: Le Seuil.
———. 1995. *La planète relationnelle*. Paris: Flammarion.
Breton, P. 1990. *Une histoire de l'informatique*, Coll. "Points". Paris: Le Seuil.
———. 1992. *L'utopie de la communication*. Paris: La Découverte.
Butera, F. 1991. *La métamorphose de l'organisation: Du château au réseau*. Paris: Les Éditions d'Organisation.
Canguilhem, G. 1993. Le cerveau et la pensée. In *Georges Canguilhem, historien des sciences*, Bibliothèque du Collège International de Philosophie. Paris: Albin Michel.

---

[23]Interview in the newspaper *Le Monde 2*, 28 May 2005.

Castells, M. 2001. *The internet galaxy*. Oxford: Oxford University Press.
———. 2010 [1996]. *The rise of the network society*, vol. I, 2nd ed. with a new preface. Chichester/West Sussex/Malden: Wiley-Blackwell.
Chevalier, M. 1836. *Lettres sur l'Amérique du Nord*, 2 vols. Paris: Gosselin.
———. 1842. *Cours d'economie politique fait au Collège de France*. Paris: Capelle.
De Kerckhove. D. 2000. *L'intelligence des réseaux*. Paris: Odile Jacob.
De Rosnay, J. 2000. *The symbiotic man: A new understanding of the organization of life and a vision of the future*. New York: McGraw-Hill.
Dery, M. 1996. *Escape velocity: Cyberculture at the end of the century*. London: Hodder & Stoughton.
Ferguson, M. 1987. *The aquarian conspiracy: Personal and social transformation in our time*, New York: J.P.Tarcher.
Flichy, P. 2002. *The internet imaginaire*. Cambridge, MA: The MIT Press.
Gibson, W. 2011 [1983]. *Neuromancer*. New York: Harper Collins.
Gille, B. 1978. *Histoire des techniques*, Coll. "La Pléiade". Paris: Gallimard.
Goody, J. 1977. *The domestication of the savage mind*. Cambridge/New York: Cambridge University Press.
Grange, J., P. Musso, P. Régnier, and F. Yonnet. (eds.). 2012. *Henri Saint-Simon: Oeuvres Complètes*. 4 vols. Paris: PUF.
Gras, A. 1997. *Les macro-systèmes techniques*, Coll. "Que Sais-Je?". Paris: PUF.
Guillerme, A. 1988. *Genèse de la notion de réseau*. Study for the Ministry of Equipment. Paris.
Hiltz, S.R., and M. Turoff. 1978. *The network nation: Human communication via computer*. New York/Reading: Addison-Wesley.
Houston, J. 2000. Cyber consciousness, *Yes,* march http://www.yesmagazine.org/issues/new-stories/cyber-consciousness.
Hughes, T.P. 1983. *Networks of power: Electrification in Western Society, 1880–1930*. London/Baltimore: The John Hopkins University Press.
Huitéma, C. 1996. *Et Dieu créa l'internet*. Paris: Eyrolles.
Kelly, K. 1995. *Out of control: The rise of neo-biological civilization*. Menlo Park: Addison-Wesley.
Kerven, G.-Y. 1993. *La culture réseau: Ethique et écologie de l'entreprise*. Paris: ESKA Editions.
Kropotkin, P. 1910 [1898]. *Champs, usines et ateliers ou l'industrie combinée avec l'agriculture et le travail cérébral avec le travail manuel,* 2nd ed. Paris: Librairie Schleicher Frères.
Lamé, G., E. Clapeyron, S. Flachat, and E. Flachat. 1832. *Vues politiques et pratiques sur les travaux publics de France*. Paris: Imprimerie d'Everat (Paris: Bibliothèque de l'Arsenal. Fonds Enfantin FE 901).
Le Breton, D. 1999. *L'Adieu au Corps*. Paris: Métailié.
Lecourt, D. 1999. *Les piètres penseurs*. Paris: Flammarion.
Legendre, P. 2001. *De la société comme texte: Linéaments d'une anthropologie dogmatique*. Paris: Fayard.
Lévy, P. 1999. *Collective intelligence: Mankind's emerging world in cyberspace*. Trans. R. Bononno. Cambridge, MA: Perseus Books.
———. 2001. *Cyberculture:* Report for the Council of Europe. Minneapolis: University of Minnesota Press.
Licklider, J. C.R. 1960. Man-computer symbiosis. *IRE Transactions on Human Factors in Electronics* HFE-1(March): 4–11.
Lyotard, J.-F. 1979. *La condition post-moderne*. Paris: Editions de Minuit (*The post-modern condition: A report in knowledge*, Manchester University Press. 1984).
McCulloch, W.S., and W. Pitts. 1943. A logical calculus of the ideas immanent in nervous activity. *Bulletin of Mathematical Biophysics* 5(4): 115–133.
McLuhan, M. 1962. *The Gutenberg galaxy*. Toronto: University of Toronto Press.
———. 2001. *Understanding media: The extensions of man*. Routledge Classics. http://beforebefore.net/80f/s11/media/mcluhan.pdf.

Mumford, L. 1934. *Technics and civilization*. Chicago: University of Chicago Press.
Musso, P. 2000. Le commanagement et les appareils idéologiques d'enterprise. *Sciences et Société* 50/51(May–October): 149–172.
———. 2003. *Critique des réseaux*. Paris: PUF.
Naisbitt, J. 1984. *Megatrends*. Milano: Sperling and Kupfer Editori.
Nora, S., and A. Minc. 1978. *L'informatisation de la société française*, coll. "Points". Paris: Le Seuil.
Pinet, G. 1898. *Écrivains et penseurs polytechniciens*. Paris: Paul Ollendorff Editeur.
Proudhon, P.-J. 1848. Organisation du crédit et de la circulation, et solution du problème social sans impôt, sans emprunt. In *Oeuvres complètes de Proudhon*. Paris: Edition Rivière.
———. 1851. Idée générale de la Révolution au XIXe siècle. In *Oeuvres complètes de Proudhon*. Paris: Edition Rivière.
———. 1855. *Des réformes à opérer dans l'exploitation des chemins de fer*. Paris: Garnier.
———. 1868a. *De la concurrence entre les chemins de fer et les voies navigables*. Paris: Editeurs Lacroix and Verboeken.
———. 1868b. *Oeuvres complètes*. Paris: Librairie Internationale.
Rifkin, J. 2000. *The age of access: The new culture of hypercapitalism*. New York: Jeremy P. Tarcher/Putnam.
Ruyer, R. 1950. *L'utopie et les utopies*. Paris: PUF.
Saint-Simon, H., and B.-P. Enfantin. 1865–1878. *Oeuvres de Saint-Simon & d'Enfantin*, 47 vols. Paris: Éditions Ernest Leroux.
Schlanger, J.-E. 1971. *Les métaphores de l'organisme*. Paris: Librairie Philosophique Jean Vrin.
Sérieyx, H., H. Azoulay, and CFC Group. 1996. *Face à la complexité, Mettez du réseau dans vos pyramides: Penser, organiser, vivre la structure en réseau*. Paris: Village Mondial.
Servan-Schreiber, J.-J. 1967. *Le défi américain*. Paris: Denoël.
Spencer, H. 1883–1890. *Principles de sociologie*, 2 vols. Paris: Germer, Baillière & Co.
Stourdzé, Y. 1998. *Les ruines du futur*. Paris: Sens & Tonka.
Théry, G. 1994. *Les autoroutes de l'information*, Report for the Prime Minister. La Documentation Française. Paris.
Toffler, A. 1985. *The adaptive corporation*. New York: McGraw-Hill.
———. 1986. *S'adapter ou périr: L'entreprise face au choc du futur*. Trans. M. Deutsch. Paris: Denoël.
Von Neumann, J. 1958. *The computer and the brain*. New Haven/London: Yale University Press.
Wiener, N. 1988 [1950]. *The human use of human beings: Cybernetics and society*. Boston: Da Capo.
Winner, L. 1986. Do artifacts have politics? In *The whale and the reactor: A search for limits in an age of high technology*, 19–39. Chicago: University of Chicago Press.

Chapter 3
Network, Utopia and Fetishism

Filipa Subtil and Pedro Xavier Mendonça

Introduction

One of the achievements of Pierre Musso's text (Chap. 2) is that it supports the argument that the concept of communication changes with the network's social imagery. From this point of view, a new conception of communication, with a large mobilizing potential that goes beyond the role of the press, began establishing itself during the nineteenth century, and had its origins in the idea of the network. By the beginning of the nineteenth century the network metaphor was replacing that of the human body and medicine for land transport routes. At the time, the network was conceived as a large, self-regulating machine superimposed on the land, with the railway becoming the best example of this. The network, which had been biological, had become technical. While the physician observed the network, the engineer was able to conceive it and to build it.

The term "communication" began to be understood as a technical system of networks under the influence of such renowned disciples of Claude Henri de Saint-Simon (who Karl Marx as labelled as "utopian socialist") as Barthélémy-Prosper Enfantin and, later, Michel Chevalier. There were two types of theoretical network: material – which were mainly identified as transportation lines; and immaterial/ "spiritual" – which were associated with the financial flows of banks. The new communications and transport networks were thought of both as mediators of social change and as producers of social relationships, and even of a social transformation capable of leading mankind to the ideal of "universal association" and to a "peaceful future of prosperity and glory" (Chevalier 1832, quoted by Musso 1999: 108).

F. Subtil (✉)
School of Communication and Media Studies, Lisbon Polytechnic Institute, Lisbon, Portugal
e-mail: fsubtil@escs.ipl.pt

P.X. Mendonça
School of Business Communication-EFAP, Lisbon, Portugal

In this chapter, we discuss the relationship between the imagery favoring these technical networks and the expectations they bring, leading to effects similar to the concept of fetish. Beginning with the thoughts of Musso and his contribution to this topic, we seek to present elements of that imagery and its technical realization in what is referred to as communication.[1] We then seek to deepen the connections Musso established with these dynamic, utopian ideologies and a certain kind of network fetishism experienced in the political component. We argue that utopian promises of the technological development of communication networks promoted fetishist experiences. We will tie them in with the tradition of the social studies of technology that describe different ways in which interaction with technology occurs, giving rise to a not fully rationalizable attraction.

Social Imagery of the Network and Its Technical Realization

In the text being discussed, as elsewhere, Musso (1997, 2003) argued that the Saint-Simonian imagery offered the substance of a Promethean technological utopia based on the idea of the communications network. The first attempt to create technical networks with which to unite spaces and dispersed communities appeared in the ideas outlined by Enfantin during the first quarter of the nineteenth century. These were the roots of the creation of a system of railways and canals supported by the banking system Chevalier was pushing for. It sought, with some tenacity, to reduce distances through a networking dynamic which depended on a physical structure.

Saint-Simon had outlined a proposal for social and political organization in which the pooling of interests resulting from the industrial system would seek to reconcile the efficiency and economic self-interest that were the hallmarks of this new scientifically and technologically-based economic organization, with higher moral sentiments, as the only way to ensure a peaceful, cohesive and prosperous community. Scientific, technological and economic progress provided an outstanding opportunity to construct a universal association in which self-interest could not exist without a common set of moral ideals. Creating a society requires more than moral and economic self-interest, however, and more than well-organized interests, it also requires a goal that can be taken as the ultimate aim of human behavior – interest in others, sympathy and moral solidarity. From here, the followers of Saint-Simon tended to identify that "universal association" with material and financial networks.

Pierre Musso and Armand Mattelart have both highlighted the role of the works of Enfantin and Chevalier. Musso notes that Enfantin, writing in *Le Producteur*

---

[1] See also Flichy (2007) who applies this connection between imagery and its technical realisation to the birth of the Internet.

(one of his movement's official publications[2]) in May 1826, argued that "the general system of communications should be applied on a global scale, the territory of man, in order to achieve the universal association seeking to 'develop the combination of efforts towards a common goal: the exploration of the world we inhabit'. Enfantin developed the idea of a combination of material transport networks and immaterial credit networks that would establish a sophisticated general communications system. *Le Producteur* published many articles on the subject of money and credit – and on immaterial exchange, which was later described as 'spiritual'" (Musso 1999: 100–101). Thus, it was the polytechnic engineer rather than the man of letters or the journalist who, through his professional activity, was to become the protagonist of the networked industrial society.

Mattelart notes that Chevalier was a critic of egalitarian socialist movements and at the same time a strident opponent of views emphasizing the evils of the railways (Mattelart 2000 [1999]). Chevalier argued strongly for the political legitimation of technical and financial communications networks. To Chevalier, in *Lettres sur l'Amérique du Nord*, improving the communication system in United States was: "working for real, positive and practical freedom; to allow all members of the human family to choose to travel and explore the world they received as their heritage; to extend the freedoms of the greatest number and to do so as often as possible through laws of exception. I will go further, it creates equality and democracy. Perfected means of transport reduce distances, not only between two points, but also between two classes" (Chevalier 1836: 39).

Chevalier's faith in the potential of networks led him to claim that the role of railways was identical to that of religion: *religare*. This mode of transport was the most powerful structure connecting scattered communities and making relationships between people peaceful, particularly those who had settled around the Mediterranean basin, that historic point of connection and confrontation between East and West. Chevalier argued that the longed-for Mediterranean confederation, created by a number of intersecting and overlapping technical networks, had to be governed by the power centers of technological communications or it would be in danger of descending into anarchy. In his view, states with technological power are the best candidates for governing and imposing their will, giving them the obligation to energize the inert peripheries surrounding them.

Thus, by the end of the nineteenth century, together with the concept of communication as a means of culture and the creation of a questioning community that originated with the literary and humanist tradition of the intellectuals, a trend developed, one associated with a more technological dimension that was linked to transport and transmission over distances. At the same time, the symbolic means of communication began witnessing a growth of commercial newspaper advertising and an accompanying reduction in the number and length of news articles. Large

---

[2]In addition to *Le Producteur*, which was published from June 1825 to the end of 1826, there was also *L'Organisateur*, which was first published in August 1829, and *Le Globe*: *Journal de la Doctrine Saint-Simonienne*, published between 1830 and 1832.

companies began to prefer engaging with their customers through newspaper adverts, transforming the newspaper into a powerful tool for achieving commercial goals. News meeting market expectations began to be exploited, and newspapers became increasingly dependent on advertising (Subtil and Garcia 2010). At the same time, the press was becoming increasingly independent of the political authorities, although it maintained close links with the political world. Its growing independence from political power encouraged the emergence of a journalism that would challenge political figures and uncover cases of corruption.[3] This enabled the press to increase its audience, which in turn attracted more advertisers. As Robert E. Park (1923: 273–289) noted, in terms of mass production, distribution and advertising, modern newspapers anticipated future forms of commerce and consumption.

In a society marked by economic growth and by the routinization of technological innovation, the information sphere takes on an extraordinary role, and this brings about profound changes. Transport improvements, changes in city lifestyles and the social division of labor encourage growth in the search for news and of many of the media's ancillary products, which are then largely supplied by the press. In these circumstances, and thanks to the previously unattainable large-scale capacity for dissemination, newspapers become powerful social institutions (Weber 1976 [1910]: 96–10; Hardt 2001: 127–141). In its role as a machine for convincing and connecting those with the ability to exercise social and political leadership, the press consistently increased its own influence over public opinion, to become the most important vehicle for the transmission of ideas between social groups, particularly between the authorities and the public, as well as for different types of propaganda and social mimicry.

A powerful and effective information tool, the press also quickly transformed into an essential pillar of the establishment of the market economy, while simultaneously assuming the form of an economic product. Its survival was now determined by commercial success, which was embodied in advertising revenue and newspaper subscriptions. Despite the possibilities opened up by information technology, the economic condition in which the press found itself had ambiguous, even devastating, consequences for reporting facts and on the quality of journalism. Based on the constant need to attract new readers while also retaining existing ones, the economic position of the press led to the production and publication of superficial and sensational news – a form of journalism focused on the present and the new, and increasingly removed from reflection and debate (Subtil and Garcia 2010).

Parallel to this, radio and television were created during the twentieth century, after the invention of cinema, as a result of the application of electronics to the media sector. With the expansion of this new broadcast media, a new and

---

[3] Perhaps the archetypical case of this type of journalism was the "muckraking" movement, a journalistic trend that unreservedly described the social, political and economic transformations affecting culture in general. It publicly denounced scandals and social injustices, concentrating on the poverty in the suburbs of the large cities, on factory working conditions, child labour, etc. For more on this type of journalism, see Emery, Emery and Roberts (2001 [1984]: 223–226) and Eksterowicz, Roberts and Clark (2003 [1998]: 91).

increasingly independent "communications management" sector appeared within companies, which included public relations, advertising, marketing and communications and information services. A second area of media communications then emerged and developed rapidly – telecommunications.[4] This involved a whole range of services and technologies for transmitting, emitting and receiving signals, written messages, images or sounds over wires, optically or by other electromagnetic systems.

The word "telegraph" appeared in 1792 with the first semaphore lines. The introduction of the telegraph created the technical conditions necessary for the growth and national expansion of this media and led to the creation of a national audience. The media then abdicated its local and regional responsibilities even as it aimed for an ever larger audience and for the first time came into direct contact with the national community. In fact, the railways and the telegraph provided the infrastructure for a national society (Carey 1997: 322–323). This new and large national market fueled the growth of telegraph services – which for financial reasons demanded a simple, short and standardized form of writing that was free of colloquialisms – to serve larger and more heterogeneous audiences. The origins of journalistic objectivity can, therefore, be found in the need to transmit messages across space through telegraph lines. This long-distance communications network contributed greatly to the growing production and spread of poorer and lower quality prose in the newsrooms. Given the exponential growth of news items arriving in the newsrooms, they found themselves having to change their organization model completely. News procedures were routinized, and the structure and organization of newsrooms increasingly came to resemble factory production lines. As with any other product, news was subject to a whole series of industrial processes. Profit-driven, large-scale news production contributed decisively to the disappearance of a certain genre of journalism. Urgency was the value that came to determine the relationship between information producers and an audience that was no longer concerned with detailed, analytical journalism (Carey 1992 [1989]: 201–230; Subtil 2014).

While in the information sphere the nineteenth century was the age of the written press, the twentieth century was one of the alliance between information transmission, abstraction and opulence that from the 1980s became an axis of the new capitalism and of a new society. This trend developed from the telegraph to the telephone, which in turn gradually developed the characteristics that led to the current information age.

From the optical telegraph and Claude Chappe's telegraph systems, through Samuel Morse's electric telegraph, Marconi's radio transmissions and Graham Bell's telephone industry, the foundations of the long-distance telecommunications industry had been laid. Throughout the period from 1935 to the beginning of the 1950s initial technological developments were prolonged, with many new

---

[4]The word "telecommunication" was first used by in 1900 by Edouard Estaunié, a telegraph engineer.

information technology discoveries driven by the war effort. The first computer, ENIAC, began operating at the University of Pennsylvania's Moore School of Engineering in 1945. Norbert Wiener, an MIT professor and the founder of cybernetics – the science of control and communications – sought to extend it to the design of human prostheses, neurophysiology and communication systems. In 1949, William Shockley, Walter Brattain and John Bardeen received the Nobel Prize in Physics for their development of the transistor, which replaced the vacuum tube as the basic component of electronic systems. At the same time in Los Alamos, New Mexico, the mathematician John von Neumann was engaged to calculate the feasibility of the plans for the H-bomb, and in particular to mathematically model the explosion, a task requiring an enormous number of calculations (one million punch cards). Multidisciplinary research in the field of information technologies, which was considered essential and, therefore, justified by military need, was developed and supported by federal funds.[5]

The third and final area was organized around computer science (informatics): the automatic processing of information.[6] Although the use of applied mathematics, and in particular the application of calculus to the technique, dated back to at least the artillerymen of Charles VIII in the fifteenth century, it was not until the twentieth century that the conditions existed through which communications could be treated as a calculation. The mathematician Alan Turing, who formalized the notion of the algorithm, made an important contribution to the theoretical foundation of modern computer science (Breton 1987; Lévy 1996 [1989]: 157–183).[7]

From the beginning the computer was developed for top-secret bellicose ends, and soon became an indispensable tool for current applications. In 1951, UNIVAC became the first civil computer to be sold to the public. During this decade the civilian market for computers was limited and insignificant compared to the needs of war and the government. Only a few dozen machines were turned over to civilian applications during this period, and the outlook for their development was not encouraging. By

---

[5]Public opinion in the United States in the wake of the global events of the 1940s was strongly supportive of scientific research, which was considered essential as a means of finding solutions to almost all problems. North American public opinion identified long-distance technologies as "technologies of freedom" (Breton 1987).

[6]This term derives from the combination of the French words *information* and *automatique*, and was coined by Philippe Dreyfus in 1962. In 1966 the French Academy recognised it with the following definition: "Science of the rational treatment, in particular by automatic machines of information considered as the support of human knowledge and communication in technical, economic and social matters".

[7]The ground for informatics was also well prepared by the development of mechanography, invented by Hermann Hollerith, founder of the company that went on to be called IBM. This technique sought to mechanise the collection and treatment of statistical and accounting data and, more generically, of all the social and economic information produced. The perforated cards used as the data medium benefited from being universal, which earned it significant – albeit temporary – success. Despite its ubiquity, mechanographic machines soon proved impractical when faced with the exponential growth of requirements.

1960, there were only a few thousand computers being used for non-military means, 500 of which were the famous IBM 650s that had been sold to the traditional mechanographic market. By 1966, there were 34,900 computers in the United States, with 10,700 of these either directly or indirectly in the service of the state. Three-quarters of those used by the government were being used by the Department of Defense. Fifteen years after the sale of the first computer, 50,000 computers, with a value of around $20 billion, had been installed across the western world. Since then the industry has grown with unprecedented speed. During this first phase the main producers were the United States, Japan, France, the United Kingdom, Germany and Italy, the only countries in the world with a Gross National Product that was sufficient to enable them to make a commitment to this new sector of industrial activity. It was not long before eight US companies came to dominate the sector, holding a 90% share of the global market between them (Breton 1987: 196–197).[8]

The emergence of micro-computing at the beginning of the 1970s radically altered both small applications and computer science as a whole. The micro-computer was invented to challenge the centralization and concentration of inside information in the hands of some groups. It was a kind of revolution within a revolution, and its radicalism is partly due to the origins of computer science culture, which was shared by a large public and was itself a factor in the democratization of social life and knowledge (Breton 1987: 228–233). To the extent that Shannon's model underlies computer science, it is now more than ever contained in explanations of the "information society", contributing, along with cybernetics, to the development of the information economy or what Dan Schiller termed "digital capitalism" (2014, 2000 [1999]). This is how the world network, dreamt by the utopian Promethean followers of Saint-Simon, finally took shape in the information highway, in many ways quite different from that which had been imagined.

The information and communications fields spilled over into the press, increasingly imposing itself as an economic sector that was important and, indeed, crucial for the production and management of market-valued knowledge. Over recent decades, this information has come to include the media, telecommunications and computer science. As was true during other periods, the convergence between these sectors has been accompanied by conceptions of a utopian nature, retaining the vision of communications as an alternative to the political disorder of the world during the inter-war years (Breton 1997 [1992]). Since the end of the 1980s, this reality has been identified and analyzed in a range of theories and studies. Breton and Proulx call the confluence of the various domains of technical and economic information the "explosion of communication", which has accompanied the "utopian project of a communications society" (1997 [1989]: 22). Similarly, the editor of this book, José Luís Garcia, dubbed this a new communications constellation in the economic plan (Garcia 2014).

However, with the popularization of the expression "communication" to cover the various areas in focus and, at the same time, the way they have acted on each

---

[8]In 1962 IBM was responsible for 65.8 % of global computer production (Breton 1987: 198).

other under the effect of the multi-level possibilities they have opened up, we must reflect on the meaning of the different traditions in which each of these sectors are rooted. What are their differences in socio-cultural terms, and what type of complex evolution is under way?

Strictly speaking, the concept of communication, despite the central place it has obtained through scientific, disciplinary, technological and operational research, never achieved precision and stability precisely because of the diversity of specialized fields of communication (Craig 2007 [1999]; Frade 1991). However, the trend to increasing integration of different information and communication technologies – and of the associated economic world – has made it more difficult to analyze and access a unified sense of the concept. This phenomenon is the result of the evolution of information transmission technologies in which they are both faced with differences that are in certain areas dissipating, just as in others they tend to follow a path of mismatches and divergence. This line of reasoning recognizes that in the appreciation of three new territories of targeted communications, it is important not to minimize the initial differences between them, nor the complex evolution between them over recent decades (Breton and Proulx 1997 [1989]: 116). We may add that several of the initial differences in communication technologies cannot be analyzed as simple historic contingencies, the effects of which will dissipate, since these discrepancies are certain to remain and grow (ibid.). The complex evolution of the current information constellation implies, on the one hand, a trend towards the homogenization of the three information territories and, on the other, the preservation of certain stubborn differences.

It can be argued that the sequence moving from electronics to telecommunications and to information is the innovation upon which the foundations of the information age rest, given their ability to transmit and process data. This movement can also be seen in two other far-reaching transformations: the move from cities to the suburbs and the creation of large industrial and commercial corporations.

If there is an exemplary case of this radical transformation of society through information, it is certainly the United States, the world's leading political, economic and cultural power. What tradition perhaps still means for Europe, information certainly means for the United States. This matter has been discussed by Chandler Jr. and Cortada: "This is a nation that enjoys the taste of information. This is more than just an appetite needed to sustain economic life. Across almost all aspects of their economic and political life, Americans have reached out to technology and built the business infrastructure necessary to deliver it to the economy, with which to solve problems or to define how things are done. It is not an accident that the United States led in the exploration of space, not just for altruistic reasons, but for practical purposes. America's love affair with the automobile is as much an infatuation with how it works as it is with freedom of movement it promises all drivers" (2000: 299).

It is not our intention to establish a predictable and linear thread linking the Saint-Simonian tradition of technology in the service of the common good, particularly of the popular and poorer classes, and the current technological path that serves its own expansion, the most powerful classes and profit. This process is long and involves many unforeseen and complex factors.

Networks, Musso and Attraction of the Political Network

From this framework, it is possible to highlight one aspect in which Musso's work is particularly significant. That is the relationship between communication networks in their political component, with utopian characteristics, and a certain non-rational attraction that Musso associates with fetishism.[9] This emphasis demonstrates a network power shift from instrumental logic to an appreciation of the technique itself as utopian consummation, an approach that establishes a link between utopian political "retiology" and the tradition of social studies of technology that project sublime or enchanting effects of technology onto individuals. This is not simply the result of a process or technical finality, but also of the political promises written into communication. The network as manifestation acquires an emanation distinguishing it as a political reality.

As we have seen, Musso discussed the links between the network as representation and material reality and some of the utopian movements of the nineteenth and twentieth centuries. The dual consideration of the imagery and the material combine with the projection of an ideal and the search for its realization. It is here that the factors of a not fully rationalizable attraction appear, many emerging from the utopian experiences that in these cases view communication technologies as a vehicle.

In his description of the way in which the network appears in these areas, Musso (see Chap. 2) states it is taken as a kind of divinity. The worship of the network by the disciples of Saint-Simon was a blend of economic, political and religious motivations. The fetishization of the technical object "the network" as a symbol of "universal association" and as a mean of achieving the social change described by Chevalier fits this field as a political trait. It is a process that was reproduced in the past through the railways and today via information and communication technologies (ICT) and the Internet, and which is revived with each innovation. It is a twofold dynamic – political-symbolic and technical-financial. However, at the level of both representation and materialization, the former is how the latter is incorporated into the fetish mode. When Enfantin opted for the religious component of this pendulum, with the network being viewed as a symbol of universal communication integrated with "mother earth", he also followed this mystical line.

According to Musso (Chap. 2), Proudhon went further, with the link between the fetishization of the network and politics, as well as with the materialization of the fetish in the technological construction. For him, representation more strongly incorporated the technical object as a political bias and, given the temporal dimension, the utopian perspectives. This means that the actual network represented a socio-political configuration through its materialization. A network may be more or

---

[9]For this concept we use the framework proposed by Schutz (1970) in terms of the different meanings of the term "rational". Here we refer to the ideas of logical thought and planning. Consequently, the not fully rationalisable falls into the sphere of the illogical because it is emotional, incalculable and unpredictable.

less centralized or distributed across agencies, these presenting differences in hierarchy and power relations, or organizations that are more or less democratic.

Musso (Chap. 2) proposed six main markers, or identification criteria, for "retiology", in which we encounter factors that show what we wish to highlight: an imagery that connects the network and the body in an isomorphic sense; a schematic formalization of the interpretation of reality; a promise of social transformation; an appeal for universal association; the hint of a new economics; and a disposition for the constitution and organization of power from these configurations. Of these, we focus on the promise of social change and universal association integrated into the material component that is the organization of power, and those markers that are more closely related to utopia, technology and not fully rationalizable attraction factors as the fetish referred to by Musso. This combination does not always respect the balance between these different indicators; moreover, it is also possible to find meanings in the other markers that can be incorporated into these aspects. Consequently, the network ideology has technical-messianic features that could shift the ideological component towards the processes of attraction in technology itself: fetishization in its real sense. Here the technique stands out as a self-referential tendency, becoming a vehicle for utopian aspirations, as if it could lead to the transformation of society on its own, a state that copes well with the globalization of the market and which sees innovation as the way forward.

The idea that the modern and western object could become a fetish is suggested, for example, by Karl Marx (1909) when he mentions the creation of the cult of merchandise that is separated from its use value, although the question of technology does not appear as the central theme as in other authors. From a perspective closer to the social studies of technology, Leo Marx (2000) notes elements of the sublime sphere in American industrialization, due to the pastoral environment to which it is associated. His disciple, David Nye (1994), explored the notion of the technological sublime in the industrial architectonic tradition. The large Victorian and American infrastructures provoked a wonder that was overwhelming and beyond rational judgement: a certain aesthetic of the magnificent marking a relationship with technological achievement. Carey and Quirk (1992 [1970]) tried to understand how the utopian impulse in America gave rise to a particular narrative: "the rhetoric of the technological sublime". They distinguished in this rhetoric a discourse of a futurist mentality that projected an America capable of adjusting technology and nature in such a way as to reverse the historic cycle of human poverty and misery. America would take wealth, power and productivity from industrialization, and from nature it would take peace, harmony and self-sufficiency. Mosco (2004) applies this same concept to modern digital technologies, where the notion of the network plays a central role. The concept of the "sublime" refers to a philosophical and aesthetic tradition of some complexity that is not always fully integrated into these approaches; nevertheless, it enables an understanding of technological expressions surpassing merely instrumental presence – even if that is where the stimuli are found – approaching the relationship with the classical aesthetic technique and often with a certain mysticism.

The concept of enchantment also fits into this area of analysis. George Ritzer (2000) refers to the re-enchantment of the world following the disenchantment noted by Max Weber in relation to secularization, via the processes of consumption, persuasion and technology that are clearly present in advertising and shopping centers. From a different point of view, the anthropologist Alfred Gell (1999), while establishing an anthropology of art, identifies spheres of technological enchantment through the labor the creator induces into the final object. This enchantment is not of the aesthetic or commercial and persuasive orders, but is rather technical-procedural – the enchantment anyone feels in relation to the technical skills inducted in a final constructed object.

Richard Stivers (2001) enhances these traits, proposing a synthesis of approaches in which there are factors of such technology, attributing magical properties to them. From this point of view, technology becomes magical because it causes effects in the transcendent or irrational spheres. Here Stivers (2001) highlights the irony: a traditionally rational or rationalist image in this domain is contradicted by non-rational, unforeseen or even mystical effects. In reality, the rational can, in a planned manner, provoke the irrational, particularly producers in relation to consumers or the authorities in respect of citizens. Authors with constructivist tendencies, such as those belonging to the actor-network school (Latour 2005), confirm the process of technological production is much more irrational and unpredictable than is generally believed.

It is in the general framework of the not fully rationalizable that we can place the specific concepts used by Musso and other authors in thinking about technology. Moreover, these ideas are integrated into the relationship Musso establishes between the network and utopian policy. The idea of fetishism cannot be reduced to those of the sublime or enchanted, and nor do these two mean the same. However, any of the three can refer to the not fully rationalizable attraction with an appeal to political mobilization, whether or not accompanied by an explicit ideology. The political imagery of the network is crossed by utopian projections that sometimes make the fetish-objects redundant. Utopia is then part of the content of the network imagery, and its technological realizations are candidates for being experienced in a fetishistic manner. Utopia represents a disruptive component in the technological imagery in the sense of the exploitation of new political possibilities (Flichy 2007). In Musso's analysis, this utopia perhaps results in a more conservative manner of delivering the network technology, precisely as a result of an object-fetish that establishes a type of perpetual, immobile and consummated end, even while riddled with the rhetoric of change and innovation.[10]

Important agents of this technical-economic dynamic were those most directly involved in the construction of technology – the engineers – whom Musso (Chap. 2)

---

[10]There is a long tradition of critical reflections contending contemporary technology is the embodiment (and at the same time an infringement of) utopia: see Zamyatin (1983 [1924]); Orwell (1961 [1949]); Huxley (1998 [1932], 2000 [1958]); Jacques Ellul (1973); Hans Jonas (1985 [1981]); Leo Marx (2000); James W. Carey (1992 [1983]). We should note there is another current of authors that embody utopia – e.g. McLuhan (1997 [1964]) and Lévy (2000).

calls the "demiurges of retiology". Authors such as Veblen (2001 [1921]), Noble (1977) and Galbraith (1973) mention the influence of this professional group on technical construction that was more or less influenced by different political projects. This responsibility is shared: politicians, managers, marketers and users also contribute to these processes. The intensity of the effects on the construction of technology varies depending upon each group's ability to influence. Different reference frameworks also create different technical trends (Flichy 2003). Consequently, the political imagery and the social structure interfere with the realization of technology: primarily the creation of networks and their attraction factors.

As many authors have recognized, the relationship of technology with politics is not a new phenomenon. It is possible to project nationalist sentiments on technology (Gellner 2006 [1983]; Leo Marx 2000; Nye 1994; Adas 2006) or to see in some technologies, such as communications, a utopian vehicle, as this can be projected onto the idea of the network and its realizations. Langdon Winner (1989) confirms the inherently political nature of any technical choice. Associating it with a utopian project and with fetishization, the enchanting or the sublime, as vehicles for social change, it is possible to add a not fully rationalizable attraction in the political-temporal context. Politics requires mediators (technicians) who bring the power of attraction in their promises to involve individuals. At times, this mobilization attaches itself to the mediator. Networked communication technologies are a good example of this.

It is through this important mediator that the ideological content of utopia is attached to the technical network. At the same time, this content gains new qualities. It moves ideology to technology as a solution, a finality, a self-sufficient project, such as a network. And it is here the fetishistic, sublime and enchanting effects make themselves felt. There are several political, economic and social movements feeding this trend that is never absolute, but which can dominate.

An example of this is the idolatry generated around innovation (Garcia 2012). "Innovation" as representation and realization emerge as an attraction device that, curiously, is one of the keys of both capitalist (as noted by Schumpeter) and liberal mobilization. Its appeal for social change through technology is the result of private initiative. But not only this: the state also produces an appeal and incentive for this dynamic, in which ICT and networks are central. The free software movements, "piracy" and internet file-sharers are part of this dynamic, learning in the cybernetic tradition the hope in technology and the idea that it – on its own as a global configuration – can be a political end, which gives it its fetishist, sublime or enchanting nature. In this sense, it has retained a link with the communitarian intentions of the early followers of Saint-Simon. The geek culture in its promised land of Silicon Valley and in some of their publications,[11] such as *Wired*, promotes these as a trend and a way of life. In these we find the network markers and the same fetishization of which Musso speaks. It is also possible to see the political-utopian factors in this way of life, even while they

---

[11] See Winner (2006).

remain fragmented or discrete. Certain utopian characteristics are part of the technological promises of this group, even within liberal ideological frameworks. For example, the idea that we must build a network that interconnects the "virtual" Internet and the material culture surrounding us, such as objects, through an "Internet of things", contains in itself a plan for social transformation. The sublime, or an enchantment, holds out the promise of another world while delivering it in the present.

These processes use rhetoric as an instrument in two forms: as a political discourse appealing to technological construction as a means of social transformation that will realize a utopian promise (for example, the plans and appeals for the development of national innovation systems); and as a materiality that encourages consumption and use as a condition of everyday life and which immediately establishes a new political frame of possibilities of action (for example, the devices that are intended to create an electronic democracy).

Currents of critical thought like post-operaism have formulated the concept of cognitive capitalism, attributing importance to the network as a mode of production and to the consequent move from commodity fetishism to the fetishism of the symbolic. Productive systems become the outcome of an organization of ever more immaterial flows, in which the more immaterial a commodity becomes, the greater its symbolic value. Hence the success of brandization processes, which go well beyond the simple commodity itself. It is in this sense that cognitive capitalism moves from commodity fetishism to the fetishism of the symbolic, of language and, ultimately, of the imaginary. This process occurs in all phases of the economy, from finance to consumption, at the same time invading individual lives well beyond the codification of working hours (Fumagalli 2007). The previous interpretation is not limited to the economy, it also extends to the domain of politics. This binary influence takes us back once more to Musso's ideas, according to which this dynamic establishes a representation and a materialization. The intersection between these determines much of our daily life, in that it stimulates the promise persuading us and the force of an information-industrial complex that delivers "solutions" for a whole world of "problems". Dynamics of this scope, which appeal to the most diverse type of mobilization and therefore of social reactions, do not dispense with whatever may make their goals more attractive.

Conclusion

In this discussion of Musso's text we demonstrated the combination of the utopian imagery of the network and its material incorporation, which led to a fetishization, sublime or enchantment effects on individuals through technology, as a political consummation of this imagery. This is not limited to the political sphere. We also demonstrated how this aggregate of techniques and fields of information and communication became one of the most significant elements in economic terms. We must also add that, as with the formation of a national society and economy, this set of technologies plays a decisive role in the existence of a globalizing dynamic, giving rise to a trans-national information space. The existence of trans-national actors and companies in the economy (major corporations, managers and global bureaucrats) is a corollary of the possibilities resulting from the technical means to establish these links. In fact, the commitment to the development of these means accompanies the globalization of economic activity. It is, therefore, unsurprising that the large economic powers are the main players in cyberspace, where the principal goal is to expand economic activities and the market. If it is correct to speak of globalization in the information domain that is because there is a global market for information with an area of influence that is concentrated within the OECD: that is, in those advanced capitalist economies in which the large trans-national media and new media groups have formed.

The current convergence of information technologies with other technical and economic sectors that emerged during the 1970s and 1980s is based on the incorporation of knowledge and is the motor of the so-called "knowledge economy". This technical-economic complex contains its own traces of the logics and ideologies of the incorporated technological spheres and their close links with the economic sectors that are involved in new forms of knowledge production and management. To this complex we can also add what Hermínio Martins calls the "techno-recreational constellation" (2005), meaning that the new information media is a vehicle for the production and configuration of entertainment, combining games, art, information, music, etc. According to Martins, information technology has made recreation increasingly technological, and technological products increasingly recreational. These fields also incorporate not fully rationalizable attraction factors. To a large extent, they feed them through advertising and marketing. As we have seen, the views of Ritzer and Stivers give an account of this. The political aspect highlighted by Musso tends to follow these economic aspects, weaving into the media, for example, political exaltations of technology with a fascination for consumption.

The brief scenario outlined here illustrates how, in only a century, the idea of the network as applied to communication has transferred its utopian, Promethean and humanist dimension from the community of feelings and the establishment of fraternal links between people, to a globalization dynamic that is supported in forms of spatial control, the large-scale trading of goods, advertising and cultural alienation. This holds other promises, which are also utopian in nature, but which crystallized in a technology that is sufficient in itself as an economic and political endeavor, the forms and expression of which appeal to such a not fully rationalizable attraction.

Clearly in this chapter we have not discussed the undoubted scientific merits of discoveries in the ICT field, nor of the certain promising aspects that may be invested in them. At a time when information has become one of the fundamental foundations of capitalism, and when information technologies have taken on a key role in the development of a globalized market society, we have a responsibility to expose what lies in the shadows.

References

Adas, M. 2006. *Dominance by design: Technological imperatives and America's civilizing mission.* Cambridge, MA: The Belknap Press of Harvard University Press.
Breton, P. 1987. *Une histoire de l'informatique.* Paris: La Découverte.
———. 1997 [1992]. *L'utopie de la communication: Le mythe du "village plánétaire".* Paris: La Découverte.
Breton, P., and S. Proulx. 1997 [1989]. *A explosão da comunicação.* Lisboa: Bizâncio.
Carey, J. W. 1992 [1983]. Technology and ideology: The case of the telegraph. In *Communication as culture: Essays on media and society*, 201–230. London: Routledge.
Carey, J.W. 1997. Afterword: The culture in question. In *James Carey: A critical reader*, ed. E.S. Munson, and C.A. Warren, 308–339. London/Minneapolis: University of Minnesota Press.
Carey, J. W., and J. Quirk. 1992 [1970]. The mythos of the electronic revolution. In *Communication as culture: Essays on media and society*, 113–141. London: Routledge.
Chandler, Jr., A., and J. W. Cortada (ed.) 2000. *A nation transformed by information: How information has shaped the United States from colonial times to the present*, Oxford/New York: Oxford University Press.
Chevalier, M. 1836. *Lettres sur l'Ámerique du Nord. Vol I.* Paris: Librairie Charles Gosselin.
Craig, R. T. 2007 [1999]. Communication theory as a field. In *Theorizing communication: Readings across traditions*, ed. R. T. Craig, and H. L. Muller, 63–98. London: Sage.
Eksterowicz, A. J., R. Roberts, and A. Clark, 2003 [1998]. Public journalism and public knowledge. *The Harvard International Journal of Press/Politics*, 3(2)(Spring): 74–95.
Ellul, J. 1973. *Les nouveaux possédés.* Paris: Arthème Fayard.
Emery, M., E. Emery, and N. L. Roberts. 2001 [1984]. *The press and America: An interpretative history.* Cambridge: Pearson.
Flichy, P. 2003. *L'Innovation technique: Récents développements en sciences sociales, vers une nouvelle théorie de l'innovation.* Paris: La Découverte.
———. 2007. *The Internet imaginaire.* Cambridge, MA: The MIT Press.
Frade, P. M. 1991. Comunicação. In *Dicionário do pensamento contemporâneo*, dir. M. M. Carrilho, 45–55. Lisboa: Publicações D. Quixote.
Fumagalli, A. 2007. *Bioeconomia e capitalismo cognitivo: Verso un nuovo paradigma di accumulazione.* Roma: Carocci Editore.
Galbraith, J.K. 1973. *O novo estado industrial.* Lisboa: Publicações Dom Quixote.
Garcia, J.L. 2012. El discurso de la innovación en tela de juicio: Tecnología, mercado y bienestar humano. *Arbor: Ciencia, Pensamiento y Cultura* 188(753): 19–30.
———. 2014. Une critique de l'économie des communications à l'aune des médias numériques. In *La contribution en ligne: Pratiques participatives à l'ère du capitalisme informationnel*, ed. S.J. Proulx, J.L. Garcia, and L. Heaton, 49–63. Quebec: Presses de l'Université du Québec.
Gell, A. 1999. *The art of anthropology: Essays and diagrams.* London: The Athlone Press.
Gellner, E. 2006 [1983]. *Nations and nationalism.* Malden/Oxford/Carlton: Blackwell.
Hardt, H. 2001. *Social theories of the press: Constituents of communication research from 1840s to 1920s.* New York/Oxford: Rowman & Littlefield Publishers.
Huxley, A. 1998 [1932]. *Brave new world.* New York: HarperCollins Publishers.
———. 2000 [1958]. *Brave new world revisited.* New York: HarperCollins Publishers.
Jonas, H. 1985 [1981]. *The imperative of responsibility: In search of an ethics for the technological age.* Chicago/London: University of Chicago Press.
Latour, B. 2005. *Reassembling the social: An introduction to actor-network-theory.* Oxford: Oxford University Press.
Lévy, P. 1996 [1989]. A invenção do computador. In *História das ciências*, ed. Michel Serres, 157–183, Lisboa: Terramar.

———. 2000. *World philosophie: Le marché, le cyberspace, la conscience*. Paris: Éditions Odile Jacob.
Martins, H. 2005. The metaphysic of information: The power and the glory of machinehood. *Res-Publica: Revista Lusófona de Ciência Política e Relações Internacionais* 1(1/2): 165–192.
Marx, K. 1909. *Capital: A critical analysis of capitalism production*. London: William Glaisher.
Marx, L. 2000. *The machine in the garden: Technology and the pastoral ideal in America*. Oxford: Oxford University Press.
Mattelart, A. 2000 [1999]. *História da utopia planetária: Da cidade profética à sociedade global*. Lisboa: Bizâncio.
McLuhan, M. 1997 [1964]. *Understanding media: The extensions of man*. Cambridge, MA, London: The MIT Press.
Mosco, V. 2004. *The digital sublime*. Cambridge, MA: The MIT Press.
Musso, P. 1997. *Télécommunications et philosophie des réseaux: La posterité paradoxale de Saint-Simon*. Paris: PUF.
———. 1999. *Saint-Simon et le Saint-Simonisme*. Paris: PUF.
———. 2003. *Critique des réseaux*. Paris: PUF.
Noble, D. 1977. *America by design: Science, technology, and the rise of corporate capitalism*. Oxford: Oxford University Press.
Nye, D.E. 1994. *American technological sublime*. Cambridge, MA: The MIT Press.
Orwell, G. 1961 [1949]. *1984*. New York/London/Auckland: Signet Classic.
Park, R.E. 1923. The natural history of the newspaper. *American Journal of Sociology*, 29(3): 273–289.
Ritzer, G. 2000. *El encanto de um mundo desencantado: Revolucione en los médios de consumo*. Barcelona: Ed. Ariel.
Schiller, D. 2000 [1999]. *Digital capitalism: Networking the global market system*. Boston: The MIT Press.
———. 2014. *Digital depression: Information technology and economic crisis*. Chicago: University of Illinois Press.
Schutz, A. 1970. *On phenomenology and social relations*, ed. Helmut R. Wagner. Chicago/London: The University of Chicago Press.
Stivers, R. 2001. *A tecnologia como magia: O triunfo do irracional*. Lisboa: Instituto Piaget.
Subtil, F. 2014. Du télégraphe à l'Internet: Les enjeux politiques liés aux technologies de l'information. In *La contribution en ligne : Pratiques participatives à l'ère du capitalisme informationnel*, ed. S. Proulx, J.L. Garcia, and L. Heaton, 115–125. Québec: Presses de l'Université du Québec.
Subtil, F., and J.L. Garcia. 2010. Communication: An inheritance of the Chicago school of social thought. In *The legacy of Chicago School of Sociology*, ed. C. Hardt, 216–243. Manchester: Midrash Publishing.
Veblen, T. 2001 [1921]. *The engineers and the price system*. Kitchener: Batoche Books.
Weber, M. 1976 [1910]. Toward a sociology of press. *Journal of Communication*, 26(3): 96–101.
Winner, L. 1989. *The whale and the reactor: A search for limits in an age of high technology*. Chicago/London: The University of Chicago Press.
———. 2006. Ascensão e queda de uma cidade tecnológica. *Análise Social* XLI(181): 1095–1103.
Zamyatin, Y. 1983 [1924]. *We*. New York: Harper Voyager.

Chapter 4
Retiology as Ideological Determinism in the Media: A Political Economy Perspective

David Fernández-Quijada

The network as a concept is extremely useful in social science research. As pointed out by van Loon, the network can be seen as "a device for organizing and conceptualizing non-linear complexity" (2006: 307). This makes it extremely useful in an area, that of the cultural and media industries, which is becoming increasingly complex due to parallel phenomena such as globalization and convergence with other industries such as telecommunications and IT intertwining with more sociologically traditional issues of power, culture and identity.

In such a context, the network can be used as a theoretical artefact for a better understanding of these industries, or it can be applied to empirical research through social network analysis (SNA). The former has hardly been used as an explanatory apparatus within the political economy of communication or other sociological traditions, which have opted to explain the cultural industries in terms of branches and segments (Zallo 1988), multimedia groups (Miguel 1993) or poles (Vesins 1997), for example. However, the high impact that the work of the network evangelist Manuel Castells (1996) has had on communication scholars has contributed to the use of the network metaphor as a tool for understanding contemporary media markets, often as a kind of new religion. The latter approach is built on solid sociological traditions, including sociometry, graph theory, structural balance, anthropology or Simmel's sociology (Requena Santos 2003). Social network analysis is no more than "an attempt to formally describe the social structure" (Molina 2001: 16). To that end, SNA goes beyond traditional social sciences approaches based on mere descriptive data of social actors and tries to unveil the relationships between them as part of a larger structure, looking for relational data. There are a few pieces of research applying SNA and authored by scholars within media studies: Chon et al. (2003) applied SNA to the convergence among cultural

D. Fernández-Quijada
Media Intelligence Service, European Broadcasting Union, Geneva, Switzerland
e-mail: fernandez.quijada@ebu.ch

industries and the telecommunications and IT sectors; Chon (2004) used it to analyze the main transnational media groups and their geographical expansion; Fernández-Quijada (2007) illustrated with SNA the structure of the independent production supply chain into the main Spanish TV networks in terms of the volume of their output; and Miller (2011) examined the network structure of coproductions in both high grossing and in highly lauded films in the US market.

In this chapter I propose to think about networks in the cultural industries, with a focus on broadcasting, from a political economy perspective, that is, from "the study of the social relations, especially the power relations, that mutually constitute the production, distribution, and consumption of resources" (Mosco 1996: 25). Relations allow the establishment of networks, so it seems natural to study these relations from a network perspective. In this sense, I will conceptualize these networks as three different objects: first, I will focus on networks as a technical infrastructure for mass media, the traditional function attributed to networks in cultural industries; second, I will expose its position as a marketplace of ideas, in what constitutes a contemporary ideological battlefield between neoliberal and progressive thinkers and practitioners; finally, I will show its paradoxical role as physical places in the era of non-tangible and de-materialized media. My goal is to organize and explain current media in the light of its increasingly networked organization. As I will argue, network is not just a modern way to organize mass media but part of their very nature since their origins.

Networks as a Natural Infrastructure of Mass Media

Communication networks are nothing new in the history of the human race, although certainly their scale and impact have grown considerably throughout this history. The origin of this growth is linked to their industrialization. This is probably the reason why the eminent technology sociologist Lewis Mumford (1998) only included media as relevant in the third era of the development of the machine, the Neotechnic, which starts just at the beginning of the twentieth century. Mills or the steam machine engines were more relevant in the two previous eras, namely the Eotechnic and the Paleotechnic.

Despite this chronology, communication networks already existed before industrialization. And throughout the development of that process, a network ideology and mythology can be observed, as pointed out by Pierre Musso in the second chapter of this book. However, the origins of this ideology lie not in the Industrial Revolution but can be traced back to a few centuries ago; due to technical constraints, the scale of this associated ideology was smaller and its pace considerably slower and, thus, its visibility more limited. What is really relevant in the era of the Industrial Revolution is the parallel development of sociology and the first theories that try to comprehend and explain social development and the increasingly important role of machines. Our perception and knowledge of the processes operated is the main difference with previous centuries. And this fact is closely related to the development of the social sciences.

In fact, networks were one of the key elements that enabled empires to persist for centuries or to fall like a house of cards. The roads built by the Roman Empire were vast transport networks for its armies but also a communication network to formally and informally spread the word about military conquests, decrees and taxes, combining both information and propaganda. The same model with higher or lower levels of complexity was applied by previous – Phoenicia, Egypt, Greece – and later empires – Spain, Portugal, the British and the French Empires – and also for economic purposes by organizations such as the Hanseatic League.

Among all the possible examples, there is one that stands out because of its longevity: the Catholic Church. In the Middle Ages, what currently can be considered the oldest institution in the world was able to create the most sophisticated information network to date, through two interrelated elements: first, a monopoly of ways of safeguarding and reproducing the written language in monasteries; secondly, the largest propaganda network, through multiple representatives covering all its targeted territory. Together with tight control and exchange of favors with political powers, the Catholic Church was able to establish a solid monopoly closely tied to political (thus, regulatory) authorities, which tried to avoid any serious competition for the incumbent beliefs operator. As can be seen, the network of the Catholic Church was (and still is) linked to other power networks and contributed to the network of power networks. This system is a long-standing one which can be better understood when thinking about the clientelistic political and media systems still in place in countries where the Catholic Church has been traditionally very powerful, such as Spain (Arboledas and Bonet 2014; Fernández-Quijada and Arboledas 2013; Papatheodorou and Machin 2003) or the Latin American countries (Hallin and Papathanassopoulos 2002), not to mention Italy (Hibberd 2008).

However, it is well known in the history of technology that companies and institutions need to adapt to technological change; otherwise, technological discontinuity may arise and defeat the dominant design (Anderson and Tushman 1990). This principle can be used to explain the impact of the Protestant Reformation on the monopoly of the Catholic Church over the beliefs, hopes and fears of millions of illiterate Europeans. Clearly, the sponsors of Protestantism used a technological innovation, namely the mechanical movable type invented by Johannes Gutenberg, to widely circulate Martin Luther's 95 Theses. Symbolically, this technological innovation represented a first and necessary step in breaking up the monopoly of the Catholic Church over the written language.

The communication networks of ancient empires preceded the development of mass media, but clearly foretold them. They were probably efficient in terms of the available technology, and they were certainly effective. Additionally, they generated a network effect: the addition of a new person to the network represented not only a benefit for that person but also a benefit for the other members, extending the reach of their information/contacts/trade/social network.

Modern communication networks, tied to current mass media, were born and developed during the different waves of the Industrial Revolution. The application of steam machines for transportation purposes allowed the railway to be born. Given the capillarity of the rail network, often the telegraph flourished along the same extension patterns and accelerating the pace at which information could be transmitted. Its application to business and the press brought the arrival of the news agencies. First Agence Havas in France (1835) and later Wolffs Telegraphisches Bureau in Germany (1849) and Reuters in the United Kingdom (1851) changed the reach and speed with which news spread and, consequently, also had a significant impact on the press.

Later on, the development of physics led to advances in mechanics and electronics and ultimately brought into being the first wireless electronic medium, nearly a century ago: radio. Its implementation with a business and politics model created a powerful but less visible broadcast network based on the hierarchical model of a one-way network, with one institution sending messages to many receivers, which soon became millions. Television came to highlight the role played by radio, but conceptually it did not represent a new network model. Television was basically built following the model previously developed by radio: nationwide monopolies in Europe and local stations joined in a national network in the USA, with a few exceptions (Bonet and Arboledas 2011). Other countries, mainly in Asia and Africa, adhered to the European public model, while the USA influenced the Latin American countries to build commercial networks.

The development and spread of digital technologies during the third Industrial Revolution not only challenged the already established media but also created some powerful alternatives through the Internet and mobile networks. While the latter are built on a totally commercial model based on private networks (which rely upon a public resource), the utopianism of Saint-Simon is represented in the hacker spirit that enabled the Internet to be born (Himanen 2001). And despite the accelerated commercialization of the Internet, this spirit remains in projects such as Wikipedia, the free software developers' communities or the free sharing spaces, for example through peer-to-peer networks (P2P).

I will use this latter case to show how the configuration of the network challenges the dominant broadcasting model put in place in the last century. In fact, P2P networks invented nothing new, but applied the model that Alexander Graham Bell had applied to the telephone over a century earlier. Later this model was copied by Arpanet, the predecessor of Internet, connecting dispersed resources into the same network without a central server for organizing network traffic. Local area networks (LAN) were also based on this model. But prior to the Internet it was never used for the mass distribution of cultural products without a hierarchy, meaning that all the members (nodes) of the network were able to contribute to it as well as to consume the products available. This system clearly echoes some social utopianism and responds to the second technical configuration of the network that Pierre Musso identifies in this volume (Chap. 2): a biopolitical vision of the reticular confronting centralized and decentralized networks. This vision challenges the business models put in place by the already established and dominant industry, in the best

Proudhonian tradition. In this sense, this confrontation between centralized and decentralized networks echoes the previous battles in the configuration of the railway and electricity infrastructures, as narrated by Musso.

Despite this democratic configuration, some caution must be exercised when analyzing P2P performance (that is, actual use), considering the level of contribution of all the members or peers connected to those networks. From its early days, the issue of the role of so-called 'free riders' arose (Adar and Huberman 2000). These are users who benefit from the network without contributing to it and account for most of the members of these networks. The basic idea behind this concept is that a typology of users can be established according to both the amount and the quality of their contribution, producing a dual typology of free riders: users who do not contribute to the network but benefit from it (who in the original study by Adar and Huberman represented two-thirds of the total number of users) and those whose contributions are of no interest to other members of the network, rather like a TV or radio programme with a zero rating or 0 % reach. These facts, then, cast doubts on the practical implementation and actual practices of this utopian vision.

An additional cautionary note needs to be sounded when talking about egalitarianism. Contrasting with traditional broadcasting networks' role as gatekeepers of information for the audiences, the discourse on digital networks highlights the lack of hierarchies. Actually, what really happens is that new hierarchies are established, and new gatekeepers arise. The new kids on the block are often global gatekeepers, making accountability more difficult from a national perspective. In this sense, Apple's iTunes or Spotify for the recording industry, Google's YouTube for video companies and Google's search engine for news providers become key filters for millions of citizens who use them as their gateways to the cultural products mainly produced by traditional players.

Paradoxically, the authorities put limited emphasis, if any, on the monopolistic situations that these dominant positions create in a supposedly free environment. Just the opposite happens with walled gardens, closed environments such as pay television networks. There, regulations on electronic programme guides (EPG) try to set a level-playing field for different TV operators, preventing the EPG operator from undermining the order and thus the findability of the different channels or establishing a fee for obtaining a prominent position on that EPG.

This utopianism is not exclusively linked to the Internet but was previously represented in the cultural industries through the alternative press – including fanzines – and community radio and television. Without the technological and economic barriers of these previous media, these utopia-like outlets are now more visible to interested citizens through the Internet. In some cases, citizens arrive at these outlets without realizing that they represent this utopian hacker spirit. In this sense, they have become transparent, which makes it easier for them to spread but at the same time deny its utopian and political value.

As can be seen, the network as a concept is not a new idea within the cultural industries. What is newer is its digital nature, which has had (and still has) a huge impact. Digitalization is the pillar of the convergence of cultural industries with other digitally based sectors, namely telecommunications and IT. Among its effects, Galloul (1997) highlights the ubiquity of content, now de-localized, de-materialized and ubiquitous, its unlimited dissemination through networks, its infinite copy potential and the transformation of social practices, leading to new intellectual configurations.

Confronted with this scenario, Miège (1992) forecast a progressive editorialization of those cultural industries producing a continuous flow of content, such as broadcasting industries, meaning that the possibility of consuming them on demand would mean the progressive loss of this continuity. This position was held by other cultural industries researchers such as Azpillaga et al. (1998), given the growing market relevance of content production rather than distribution.

Broadcasting today is characterized by high production and low reproduction costs. This means that investment is mainly allocated to the production of an original work or prototype, while the distribution cost is not affected by the number of viewers or listeners once the technical network is operational. As new networks basically impact the distribution of the content produced, it seems more appropriate to sustain that cultural industries are actually becoming more fluid thanks to the de-materialization of their traditional technical formats. Obviously, this de-materialization is easier for some cultural products than others, but it has affected all of those using a non-broadcasting distribution model: books, records, films, videogames and newspapers. The broadcasting model has demonstrated its strength, since for the time being it is more cost-effective than IP-based networks such as the Internet or mobile networks (Friedl et al. 2014). Paradoxically, then, of all the traditional cultural industries, the broadcast industries have best resisted the competition from new Internet-based models and players.

Apple's iTunes is a major example of the disruption faced by some of these industries, where new players have focused their efforts primarily on the distribution stage, clearly showing where the weakest link was to be found. Only secondarily, and not in all cases, have these actors attempted upward vertical integration towards production once they have a foothold in the distribution of content. Netflix is probably the most recent and successful example, despite its still limited production arm.

Networks as a Marketplace of Ideas

The role of networks within the cultural industries is not just related to technical aspects. Together with its primary function as an infrastructure for distributing content widely to audiences, there is a secondary one related to its societal and cultural impact. This is what we call a marketplace of ideas, where contemporary neoliberal and progressive thinkers and practitioners compare their thoughts and visions about which cultural products should be exchanged and how it should be done.

Like every marketplace, the principle behind a fair marketplace of ideas is that all the players have equal opportunities and, consequently, there is a level playing field. This vision is defended by neoliberals, who argue that the state should not intervene in the cultural industries in any way, whether as regulator, funder or player.

In contrast, leftist thinkers and practitioners suggest that conditions for the players in these marketplaces are different. These differences would justify the intervention of the state in its different forms.

This traditional discussion in the cultural industries has intensified in recent years – especially around broadcasting –, since the arrival of new digital networks, in a context of what may be described as network determinism, a kind of technological determinism in relation to communication networks in the switchover from analogue to digital broadcasting.

In this sense, we should first take into account the large number of new networks which do not succeed in the market or which do so for a limited period of time. Digital terrestrial radio in the past decade is an example of a blind faith in digital networks without investing effort and resources in creating what is really appealing for listeners, the programmes. MySpace may represent an ephemeral success and an even more spectacular decline in social audio networking.

Additionally, even if these networks achieve market and/or social success, they do not always do so at the expense of the established players. This is the case of the leading streaming music service Spotify in Sweden, where research has shown that Spotify users are also heavy listeners to broadcast radio (Carlsson 2013), suggesting that competing platforms may be mutually reinforcing in this context.

Unsurprisingly, economic and political discourse on new media technologies tends to be uncritical and optimistic, as is the discourse in the media themselves. Often new media technologies are seen as drivers of social change without realizing that they embody the values, beliefs, hopes and dreams of the members of the society where they were developed. And their shape and functions are the result of the interplay of several stakeholders (researchers, designers, producers, marketers, distributors, retailers, users, etc.) through conflicts and agreements to solve those conflicts. As history demonstrates, artefacts have politics (Winner 1980).

In this context, a major change operates in broadcasting. Traditionally, the scarcity of radio spectrum – which is owned by states and has historically been the main distribution network for electronic media – had been a major technical justification for the different licensing systems put in place by states. At the same time, especially within the European tradition of public service broadcasting, one or several public broadcasters have played an important role within the industry in offering a public broadcasting service benefiting society as a whole, but also groups which deserve specific attention, such as children, the elderly, people with disabilities, minorities, etc. This contrasts with privately owned companies pursuing a (legitimate) economic benefit for their shareholders. In some cases, some of these private companies also had some partial public service requirements in their licenses – typically a minimum amount of specific content, such as news or locally/regionally oriented programmes – and thus were offering benefits to the whole of society.

However, the pressure on authorities to reduce or abolish standards – or make them flexible, in commercial terms – has also had an effect. An example is that of digital terrestrial radio in Norway, where public service requirements have been abolished for commercial players, who used to have them in their analogue licenses. NRK, the public service broadcaster, is now the only organization with an obligation to deliver public service oriented content. While this reinforces its role as sole provider of this socially necessary service, it is worth asking if citizens are better served in this way.

Liberalization of broadcasting industries was encouraged in most Western countries prior to digitalization, but has been intensified by it. Although theoretically it was an evidence-based policy, several authors have demonstrated how ideology can drive a process of this kind (Freedman 2008; Künzler 2012). The basic goal pursued in Europe has been to build a dual broadcasting system including public service broadcasters and privately owned commercial players. The balanced scenario designed in the 80s and 90s of the past century was counterbalanced in favor of commercial players during the switchover to digital television, and by threats to repeat it during the still unclear switchover from analogue to digital radio. This process was conducted without a prior and informed debate on the role of public and commercial media. In this case, ideology was imposed by neoliberal politicians through the discourse of the brave new world supposedly brought by digital networks.

Boosted by technological innovation and the arrival of new networks, the traditional justification for public service media (PSM) and public service content has been questioned. Ironically, when some kind of social glue seems more necessary than ever in a fragmented media and social landscape, the media that most clearly can fill this gap are questioned. Although it might have been be expected that this would have been the result of a debate about changes in society and the new challenges that this brings, the truth is that arguments have been mainly associated with the availability of new technical networks and devices, in a technologically deterministic argument. Neoliberal thinkers tend to state that the Internet and social media will provide by themselves.

From the perspective of public service media organizations, the challenges of the digital era and their necessary reaction have been summarized as follows (EBU n.d.):

- Better understand the audiences, expanding knowledge of them
- Increase engagement, promoting diversity in its different forms and setting target groups
- Set portfolio priorities, maximizing PSM distinctiveness
- Be the most relevant and trusted source of information, transforming journalistic services and adopting an *Internet first strategy*
- Become more relevant to younger audiences, developing products catering to this group's lifestyles
- Empower, curate and share, fostering cooperation with the so-called creative industries

- Accelerate innovation and development
- Create a roadmap for multiplatform production and delivery, intensifying collaboration among PSM organizations
- Ensure prominence, redesigning distribution strategies
- Transform organizational culture and leadership, improving corporate culture and governance
- Make the case for PSM, measuring their return on society

The reader will undoubtedly have noticed the technological determinism of an *'Internet first' strategy*, which is part of the increasing platformism. I propose to define platformism as the network-driven approach developed within the broadcasting organizations as a response to the challenge posed by the continuous development of new networks and devices capable of receiving their services. However, this platformism becomes increasingly difficult to justify when facts are revealed. In 2013, TV-related mobile Internet consumption in the UK – one of the most developed digital markets – reached an average of three and a half minutes versus four hours for the traditional TV set (Sweney 2014).

While nobody seems to question the use by commercial companies of a publicly owned resource for profit-making purposes, critics of public service intervention in the market question the role of public service media organizations.

Many scholars have dealt with the functions that public service must fulfil in a technologically dense environment where distribution platforms multiply available supply and fragment audiences. In this context, new functions have been added to the traditional ones attributed to public service media such as the defense of pluralism, the guarantee of the democracy or the fostering of national cultures and identities (Prado and Fernández 2006):

- Play a role of mediation and credibility among the increasing number of traditional media and new sources;
- Guarantee universal access for every citizen to important information and major communication products;
- Produce information that is socially necessary and that the market will never provide because of its lack of economic viability;
- Balance and curb new communication oligopolies, which tend to concentration and thus may reduce the desired pluralism of their contents;
- Become a pro-active force in the process of convergence between the communication sector and other social sectors, such as culture, education and health, social welfare, expanding their communication activity beyond television to respond as multimedia communication institutions.

To fulfil these objectives, public broadcasting needs to be viewed, listened to, read and shared by a large enough audience to be able to make a social impact, to influence its competitors by mastering innovation and good professional practices and to justify the investment that it receives.

In this battlefield, the forces that shaped the currently dominant neoliberal policies also propel the debate about the limits of public service media. Jakubowicz

(2006) offers three possible responses to the question of what is legitimate for PSM nowadays as far as new technologies are concerned:

- Nothing is legitimate: under this model, public service organizations are seen as having no legitimacy or even as being illegitimate. This approach is based on the belief that citizens no longer want what these organizations offer; the free market could in any event provide everything they would like to access. Consequently, the role of public service media is no longer necessary;
- Attrition model: this model only considers the legitimacy of a limited range of carefully selected services and platforms, which would see public broadcasting being converted into niche broadcasting. Obviously, these would be non-economically viable services for commercial organizations. This model is about converting public service media into a marginal institution, without social relevance and having no impact on the public sphere;
- Everything is legitimate: under this model the principle that guides public service decisions is to become an active player in every network which could help these organizations to better achieve their remit in an effective and relevant manner for the public.

Additionally, there is a second battlefield where a common threat affects both public and private broadcasters: the attempt to expel them from their traditional terrestrial platform in order to re-use those frequencies for new telecommunications services, basically mobile Internet access, creating a dispute between different networks to partially provide services of similar characteristics.

This and other external threats have also achieved what once seemed impossible: the building of ties between public and commercial broadcasters. And technical networks have been a privileged space for this collaboration. In this sense, the idea of "compete on content, collaborate in technology" has emerged as a new mantra for dealing with technical innovations and, especially, digital distribution networks. Again digital radio provides examples of this approach in countries such as Norway, Switzerland and the United Kingdom, where joint efforts can be found in the design and promotion of digital transmitters or in the launch of common Internet radio players (EBU 2014).

The players involved in this second battlefield are basically three: the policy-makers at national and international level, where the frequencies are allocated to specific services, the broadcasters and the telecommunications operators.

The policy-makers have the ultimate decision about what uses are attributed to radio spectrum. Supposedly, their decisions are based on technical considerations and the common interest. Given the impossibility of establishing frontiers in the radio spectrum and the need to develop economies of scale in the manufacturing of devices, the uses given to each segment of radio frequencies need to be agreed at international level in the World Radiocommunication Conferences (WRC) held by the International Telecommunication Union (ITU).

The second interested party, namely broadcasters, tend to join together to defend a common position. Basically, in the last WRC they felt threatened by the attempts to allocate frequencies currently used by their services to other uses and, consequently, to other players.

Finally, telecommunications operators are mainly privately owned companies (and thus profit-seeking) using a public resource for this purpose. In Europe their strength is based in most cases on their historical condition as a public monopoly. They promote a widespread discourse of technological euphoria, thanks to their substantial investments in public relations and marketing, as some of them are among the biggest advertising investors in nearly every national market. Their arguments tend to highlight the benefits of innovation per se and to blur the idea of public service media. And the World Radiocommunication Conference held in 2015 showcased the fight to use the 700MHz band between broadcasters and telecommunications operators.

The efficient use of radio spectrum – a key idea of communication policies – has to guarantee the safeguarding of this public good. Its use has to contribute to the social debate and progress as well as to generate wealth, prosperity and welfare, the traditional and undisguised aim of policy approaches to this object. Up until now, for example, television has been displaced from its historical terrestrial platform due to the fact that its digital birth has more to do with the frequencies it can release rather than a real social need.

Marginalization of the traditional broadcasting model of communication – where public service values are paramount – would pose a threat to a model conceived as a key tool for the promotion of social cohesion and a guarantee of pluralism and the strengthening of democratic coexistence. And this nature relies on the fact that the spectrum is mainly used for free-to-air delivery of this content, based on the principles of universal service. This means that, at least in Europe, these services are available to nearly all the population of a given country. In contrast, frequencies cleared from television services are assigned to telecommunications operators offering exclusively commercial services, which require direct payment from the user.

At this point of the debate, it is necessary to remind ourselves that broadcast networks are the main platform for television and radio services, even in markets where online and mobile services are well developed. Broadcast is the only free-to-air, easy-to-use platform that is available everywhere. Its coverage still outpaces mobile networks and will do so for many years to come, not least because of the strength of its signal.

There are practical efficiency issues at stake in connection with this philosophy. Broadcast networks have demonstrated their efficiency for many decades. Their technical one-way configuration has shortcomings in terms of direct audience participation through the same networks, although hybrid solutions combining two or more networks can be implemented. While broadcasters argue for the benefits of this model, telecommunications operators would like to use these frequencies, which would offer them the possibility of expanding their coverage and reduce their costs, thanks to the nature of the UHF (ultra-high frequencies) band of the spectrum currently used by broadcasters. However, telecommunications networks are configured as unicast or multicast systems. This allows them to have two-way/interactive communication but, at the same time, reduces the efficiency of delivering data to large audiences. In some cases, the costs and bandwidth needs of delivering traditional broadcast content to large audiences would be unaffordable for broadcasters and would saturate the telecommunications networks (Friedl et al. 2014; Teracom 2013). It is not about keeping broadcast at any price, which could otherwise be labelled as determinism, but about analyzing all the options and making informed decisions. In this sense, broadcast is still seen by the television and radio industry as the backbone of present and future services, which are becoming increasingly hybrid. There is no longer any room for purely broadcast strategies: broadcasting is cost-efficient for the broadcaster and for the listener, but broadband can add a degree of sophistication and interaction currently missing in traditional broadcast networks.

Networks as Places

Around the common discourse about how IT networks de-territorialize media and information services production, Pierre Musso has offered some specific critical insights. For example, Musso (2010) summarizes the "simplistic" approaches to the new territoriality defined by ICTs:

- ICTs reduce the need for journeys and trips;
- ICTs enable the de-territorialization of traditional spaces;
- A new non-tangible networked territory is created;
- ICTs allow for long-distance exchange and commerce;
- ICTs enable the de-localization of all kind of activities.

At the same time, Musso highlights that territory has not only a physical dimension but also a cultural and a political one. If we agree that networks are a new form of territoriality, this means that these networks can also be analyzed from these three dimensions.

From the perspective of the cultural industries, this distinction has profound implications. In fact, these three dimensions are loosely inter-related in most of the cultural industries, and this encourages their mutation.

First of all, some cultural industries have a physical dimension, in that they exist as a manufactured product in the form of newspapers, magazines, cinema reels, audio and video discs or any other format. The impact of networked ICTs on this dimension has been pointed out earlier in this chapter when considering networks as a natural infrastructure of mass media.

Second, the cultural dimension of networks is underpinned by the symbolic value of the content cultural industries deliver to citizens. This is their main asset: a product charged with cultural values which offer a representation of the world based on a set of ideas and beliefs. Even if these creations are embedded in processes of industrialization and marketization, they do so in a very specific way

(Huet et al. 1978), which always implies maintaining a minimum degree of symbolic input (Zallo 1992).

Finally, cultural industries also carry a political dimension through their products. Ideology is the other side of culture: if culture is a way of representing the world, politics is the way power is distributed among social stakeholders. Obviously, key issues for a democratic society arise at this point: diversity, pluralism, universality, etc.

Although networks are normally associated with distribution, in the political economy of communication they can also be seen as production sites. In its early days, radio and especially television were developed by companies covering all the stages of the value chain, from the production of the content to technical distribution. However, in recent decades broadcasters have faced a progressive process of concentration at the stage where they can add more value. This means that many activities have been outsourced, including the production of content, giving rise to a flourishing independent production industry.

Although Musso (2008) highlights the ambiguity of the digital territory, in the era of non-tangible and de-materialized media, physical place has become something very relevant for explaining current production practices. This means that despite all the de-localization rhetoric, the territory in which a media company creates and produces its content is relevant; a specific place adds value to the production, and it is often difficult to find this value in a different place, including creative workers. Thus the dramatic de-localization of some industries in the Western countries is difficult to replicate for the production of culturally loaded media productions. And this also explains their increasing attractiveness for local politicians.

The industrial practices known as clustering are probably one of the best examples of this paradoxical trend. In the economic literature, the term cluster appears closely tied to that of industrial district and, in fact, this terminological difference has also given rise to two different conceptions of how groups of companies function. The idea of the cluster, in its spatial sense, has become popular since the 1990s principally through the work of the North American economist Michael E. Porter (1990). The notion of industrial district, on the other hand, is based on the work of Giacomo Becattini (1979), who reclaims and updates the postulates of British academic Alfred Marshall (1963 [1890]).

In the industrial world, a cluster is defined as a concentration of companies related to each other, in a relatively defined geographical zone, so that they make up, by themselves, a specialized production pole with competitive advantages. This is what can be inferred from the classic definition of Michael E. Porter (1998), which establishes that they are geographic concentrations of interconnected companies, specialized suppliers, service providers, companies in related industries and associated institutions (for example, universities, standards-setting agencies and trade associations) in particular fields, which both compete and collaborate. Companies are interconnected thanks to different types of networks. Porter's central argument is that both institutional relations and relations between companies encourage innovation, international competitiveness and the capacity to grow.

Many of the industrial policies on regional and local development for developed countries over the last few decades have been based on this idea.

Public policies for clusters were created to support strategic groups of companies in a specific territory. In other words, based on the detection of a group of strong industries in a certain area, the aim of public policy was to stimulate this and encourage it via different measures, such as consultancy, research and development (R&D) and assistance with internationalization. These initial policies for natural clusters were followed by a policy of what Pacheco-Vega (2007) calls induced clusters; i.e. industries that, given their size, cannot be considered as a cluster, but where the potential has been identified for them to become a cluster in the future. And that is why they are aided.

In the case of broadcasting, British clusters are surely the best documented in the European context. Chapain and Comunian (2010) compare the cases of Birmingham and Newcastle-Gateshead and find that their potential comes from the characteristics of the city per se, while the main problems are shared, such as limited markets, a negative image and competition with the capital, London. They also highlight the need to analyze these areas' connections with other production centers to understand the phenomenon in all its complexity. For their part, Cook and Pandit (2007) chose London, Bristol and Glasgow for their comparison. Among their conclusions, they question the regional scale in the process of clustering, as well as arguing that there is little evidence that an active public policy can boost this. The Bristol study by Bassett et al. (2002) highlights the fact that smaller cities can play a leading role in the sphere of cultural production if they have the capacity to retain some autonomy with respect to the large centers, such as London. This capacity depends on some specifically local circumstances that cannot always be replicated. In the case of Bristol, public-private collaboration has been crucial to attracting investment and infrastructure. On the other hand, there are factors that go beyond the local sphere and therefore escape the action of local public policies, such as international distribution networks, market volatility and technological change. These specialized clusters, therefore, can only survive by continuous adaptation, innovation and creativity, bringing new areas of activity into the cluster.

A few other authors claim that policies aimed at promoting specific industries so that they develop into clusters are not always effective because the regions where they are applied suffer from a lack of suitable resources or the necessary institutional structures. The study by Turok (2003) on the cluster policies of the sector in Scotland points out that cluster policies *à la Porter*, focusing on developing local supply chains and local collaborative networks, are not enough to create and develop clusters that promote regional objectives. They are often even unnecessary policies. Another habitual criticism of Porter's model has been its ahistorical nature, i.e. that it defines a series of parameters that may ignore, for example, the reasons for the historical location of a firm in a particular place. Similarly, it has also been criticized for placing too much emphasis on local relations, while the importance of national and international connections has been ignored or undervalued for these groups of firms.

In short, clusters are built up on the strengths of the region itself in the form of clients and suppliers, infrastructures and natural resources, human resources available, reduced transaction costs due to limited distances, research and training centers and universities. For a balanced analysis, however, we must also evaluate the possible negative effects of proximity, such as predatory behavior in the search for clients and highly specialized labor.

As can be seen, geography still matters in the cultural industries; not only traditional hubs of creativity and production have kept their key role in attracting investment and creative minds but they have also expanded into new areas such as digital production for the new de-materialized networks.

This argument is relevant and brings a new dimension to the other arguments developed by Musso (2008) around the ambiguity of the digital territory; in the era of non-tangible and de-materialized media, the physical place has become something very relevant for explaining current production practices.

An additional argument can be added to describe the complexity of the digital territory proposed by Musso: its physicality. Despite the discourse of de-materialization and the virtual nature of communication networks, they still rely on physical components such as wires, satellites, transmitters and servers. And these components are built upon the electricity network, which is a very physical one, with a high impact on the territory (Walsh 2013).

Networks Have Ideology

As has been shown throughout this chapter, the network as a concept can shed light on contemporary broadcasting industries, not just networks as technical elements, as they have traditionally been considered, but also as places where power relations are produced and reproduced. It is a matter of connectedness: connected people, connected business on a demand and supply basis and connected ideas and thoughts.

In these networks, the degree of connectedness and the relevance of these connections explain the power position that each node has. For example, in the network of broadcasting production, broadcasters have a more central role, since their number is limited, while the number of independent production companies trying to sell them programmes is always much higher. This is a well-known phenomenon in economy, oligopsony, where sellers outnumber buyers and thus have limited power. Free riders constitute another clear example of the different value of the nodes of a network.

Naturally, it is in their configuration as a marketplace of ideas that networks best illustrate the struggle for power among the different players involved. It is not just about traditional media players trying to develop strategies in a new environment but also outside companies, which seem to find in these networks a natural place for extending their core business. Google, Apple, Microsoft and Amazon are just a few examples of these new competitors for traditional broadcasters. All these are deep-pocketed technological companies accessing the content market in different ways. Broadcasters, whose expertise is in the creation of appealing content, represent just the opposite. As content is broadcasters' main strength, it also explains why broadcasters come together to demand open networks, like those where they have traditionally operated. And they back their argument with ideas such as freedom, openness and universality.

As can be seen, the rationalist dream of a value-neutral communication space is far from materializing in broadcasting and digital networks. Communication networks carry ideological and political interest. This was clearly visible with traditional media and their editorial guidelines. However, it is not always so clear for the average citizen in the case of new networks derived from technological companies that claim to be ideologically and politically neutral.

Within this context, public policies tend to blur and disappear. As digital networks have the power to provide by themselves, according to neoliberal thinkers, the role of the state is no longer necessary, as digital becomes a rational myth linked to modernity, future, progress, economic development and social and cultural innovation (Musso 2008). Attached to this rhetoric, the state adopts a light approach, based on less direct intervention and a new role as facilitator. In the field of territorial policies, the concern here is, as pointed out by Musso (2008), the shift from principles of equality and equity to the attractiveness of territory to investors and commercial enterprises. A question that arises in this context is: why does the state renounce to its legitimate role and its capacity to act?

In this sense, Musso's argument in the chapter of this book on the ideological and technical determinism of contemporary *retiology*'s rhetoric can be clearly seen in the practices of the cultural industries. However, from my point of view, this is not something new, although it has undeniably intensified since the appearance of digital networks. There has been a shift from the welfare discourse of the post-world war period, based on an ideological consensus across the political spectrum, to the neoliberal imperatives imposed by rightist thought with little social sensitivity. The most worrying trend in these developments is the refusal to use public policies to balance the benefit-seeking strategies of commercial players. Worryingly, this goes against the very conception of the *res publica* and the common good originally sought with communication policies.

New technical infrastructures and their related business, then, represent just a step further in the capitalist history of production and reproduction, but also of its possible alternatives. And despite the fact that they share the same network, the possibilities of reaching their targets are very different; power and influence are here unevenly distributed. Networks open the field for new entrants and new business models, but the ideological configuration behind the operations developed there are just the same as they used to be. Thus traditional explanations and theories in social sciences research are still useful for explaining the processes underlying these networks and the power relations of those who control them. It is just a matter of avoiding the new and widely accepted uncritical techno-messianic narrative.

Disclaimer The author currently works for the European Broadcasting Union (EBU), which is the largest professional association of national broadcasters in the world. While focused on public service media, the EBU also carries out commercial activities. Among others, these activities include the Eurovision Network, a dedicated satellite and fiber system covering nearly all the countries and the largest in the world directly connected to broadcasters. This allows the organization to operate Eurovision and Euroradio services, rolling exchanges of media content coverage of news, sports and events all over the world.

References

Adar, E., and B.A. Huberman. 2000. Free riding on Gnutella. *First Monday* 5(10). http://firstmonday.org/ojs/index.php/fm/article/view/792/701. Accessed 16 Jan 2015.
Anderson, P., and M.L. Tushman. 1990. Technological discontinuities and dominant designs: A cyclical model of technological change. *Administrative Science Quarterly* 35(4): 604–633.
Arboledas, L., and M. Bonet. 2014. Radio in Spain. European appearance, Franco's legacy. *Javnost – The Public* 21(4): 67–82.
Azpillaga, P., J.C. Miguel, and R. Zallo. 1998. Las industrias culturales en la economía informacional. *Zer* 5: 53–74.
Bassett, K., R. Griffiths, and I. Smith. 2002. Cultural industries, cultural clusters and the city: The example of natural history film-making in Bristol. *Geoforum* 33(2): 165–177.
Becattini, G. 1979. Dal settore industriale al distretto industriale. *cj* 1: 1–8.
Bonet, M., and L. Arboledas. 2011. The European exception: Historical evolution of Spanish radio as a cultural industry. *Media International Australia incorporating Culture and Policy Journal* 141: 38–48.
Carlsson, U. (ed). 2013. *Nordicom-Sveriges Mediebarometer 2012*. Göteborg: Nordicom Sverige.
Castells, M. 1996. *The information age: Economy, society & culture. Vol. 1: The rise of the network society*. Cambridge, MA: Blackwell.
Chapain, C.A., and R. Comunian. 2010. Enabling and inhibiting the creative economy: The role of the local and regional dimensions in England. *Regional Studies* 44(6): 717–734.
Chon, B.S. 2004. The dual structure of global networks in the entertainment industry: Interorganizational linkage and geographical dispersion. *The International Journal on Media Management* 6(3–4): 194–206.
Chon, B.S., J.H. Choi, G.A. Barnett, J.A. Danowski, and S.-H. Joo. 2003. A structural analysis of media convergence: Cross-industry mergers and acquisitions in the information industries. *Journal of Media Economics* 16(3): 141–157.
Cook, G.A.S., and N.R. Pandit. 2007. Service industry clustering: A comparison of broadcasting in three city-regions. *The Service Industries Journal* 27(4): 453–469.
EBU. n.d. *Vision2020: Connecting to a networked society*. Le Grand-Saconnex: European Broadcasting Union.
EBU. 2014. *Digital radio toolkit: Key factors in the deployment of digital radio*. Le Grand-Saconnex: European Broadcasting Union.
Fernández-Quijada, D. 2007. *Las industrias culturales ante el cambio digital. Propuesta metodológica y análisis de caso de la televisión en España*, PhD dissertation, [Bellaterra]: Universitat Autònoma de Barcelona. http://www.tdx.cat/bitstream/handle/10803/4143/dfq1de1.pdf. Accessed 28 Nov 2014.
Fernández-Quijada, D., and L. Arboledas. 2013. The clientelistic nature of television policies in Democratic Spain. *Mass Communication and Society* 16(2): 200–221.
Freedman, D. 2008. *The politics of media policy*. Cambridge: Polity.
Friedl, G., P. Schäfer, and C. Scheubel. 2014. *Broadcast or broadband? On the future of terrestrial radio supply*. Munich: Bayerische Landeszentrale für neue Medien/Bayerischer Rundfunk

http://www3.ebu.ch/files/live/sites/ebu/files/Advocacy/Digital%20Radio/Broadcast%20or%20Broadband.pdf. Accessed 12 Jan 2015.

Galloul, M. 1997. Les industries culturelles contre le droit d'auteur. *Sciences de la Societé* 40: 177–193.

Hallin, D.C., and S. Papathanassopoulos. 2002. Political clientelism and the media: Southern Europe and Latin America in comparative perspective. *Media, Culture & Society* 24(2): 175–195.

Hibberd, M. 2008. *The media in Italy*. Maidenhead: Open University Press.

Himanen, P. 2001. *The hacker ethic and the spirit of the information age*. New York: Random House.

Huet, A., J. Ion, A. Lefèbvre, B. Miège, and R. Péron. 1978. *Capitalisme et industries culturelles*. Grenoble: Presses Universitaires de Grenoble.

Jakubowicz, K. 2006. *PSB: The beginning of the end, or a new beginning in the 21st century?* Paper presented at RIPE@2006 Conference, Amsterdam/Hilversum, November. http://yle.fi/ripe/Keynotes/Jakubowicz_KeynotePaper.pdf. Accessed 16 Dec 2014.

Künzler, M. 2012. 'It's the Idea, Stupid!' How ideas challenge broadcasting liberalization. In *Trends in communication policy research: New theories, methods and subjects*, ed. N. Just, and M. Puppis, 55–74. Bristol: Intellect.

Marshall, A. 1963 [1890]. *Principios de economía*, 4th ed. Madrid: Aguilar.

Miège, B. 1992. Las industrias de la cultura y de la información: Conflicto con los nuevos medios de comunicación. *Telos* 29: 13–22.

Miguel, J.C. 1993. *Los grupos multimedia: Estructuras y estrategias en los medios europeos*. Barcelona: Bosch.

Miller, J.L. 2011. Producing quality: A social network analysis of coproduction relationships in high grossing versus highly lauded films in the U.S. Market. *International Journal of Communication* 5. http://ijoc.org/index.php/ijoc/article/viewFile/896/580. Accessed 13 Jan 2015.

Molina, J.L. 2001. *El análisis de redes sociales: Una introducción*. Barcelona: Bellaterra.

Mosco, V. 1996. *The political economy of communication*. London: Sage.

Mumford, L. 1998. *Técnica y civilización*. Madrid: Alianza.

Musso, P. 2008. Territoires numériques. *Médium* 15: 25–38.

———. 2010. Le Web: nouveau territoire et vieux concepts. *Annales des Mines: Réalités industrielles* 104(4): 75–83.

Pacheco-Vega, R. 2007. Una crítica al paradigma de desarrollo regional mediante clusters industriales forzados. *Estudios Sociológicos* XXV(3): 683–707.

Papatheodorou, F., and D. Machin. 2003. The umbilical cord that was never cut. The postdictatorial intimacy between the political elite and the mass media in Greece and Spain. *European Journal of Communication* 18(1): 31–54.

Porter, M.E. 1990. *The competitive advantage of nations*. London: The MacMillan Press.

———. 1998. Clusters and the new economics of competition. *Harvard Business Review* 76(6): 77–90.

Prado, E., and D. Fernández. 2006. The role of public service broadcasters in the era of convergence. A case study of televisió de Catalunya. *Communications & Strategies* 62: 49–69.

Requena Santos, F. 2003. Orígenes sociales del análisis de redes. In *Análisis de redes sociales. Orígenes, teorías y aplicaciones*, ed. F. Requena Santos, 3–12. Madrid: Centro de Investigaciones Sociológicas.

Sweney, M. 2014. TV viewing figures show Brits prefer traditional sets over smartphones. *The Guardian*, 17 February. http://www.theguardian.com/tv-and-radio/2014/feb/17/tv-viewing-figures-brits-television-sets-over-smartphones. Accessed 11 Dec 2011.

Teracom. 2013. *Teracom White Paper: Can the cellular networks cope with linear radio broadcasting?* [n.p.]: Teracom.

Turok, I. 2003. Cities, clusters and creative industries: The case of film and television in Scotland. *European Planning Studies* 11(5): 549–565.

Van Loon, J. 2006. Network. *Theory, Culture & Society* 23(2–3): 307–322.

Vesins, H. 1997. Stratégies des groupes-médias français: Des configurations inédites. *Sciences de la Société* 40: 109–128.

Walsh, B. 2013. The surprisingly large energy footprint of the digital economy. *Time*, 14 August. http://science.time.com/2013/08/14/power-drain-the-digital-cloud-is-using-more-energy-than-you-think/. Accessed 14 Jan 2015.

Winner, L. 1980. Do artifacts have politics? *Daedalus* 109(1): 121–136.

Zallo, R. 1988. *Economía de la comunicación y la cultura*. Torrejón de Ardoz: Akal.

———. 1992. *El mercado de la cultura: Estructura económica y política de la comunicación*. Donostia: Tercera Prensa.

Chapter 5
History, Philosophy, and Actuality of the Utopian View of Technology: On Pierre Musso's Critique of Network Ideology

Steven Dorrestijn

Introduction

In this article I will explore the importance of the utopian conception of technology by looking into the history of technical utopias, its significance in the philosophical understanding of technology, and its relevance in current debates. If "technical mediation" is considered a general term for the ways in which technology influences and co-constitutes human existence, then this essay is about the optimistic, utopian outlook on this subject. As this utopian view is somewhat neglected in the field of contemporary philosophy of technology, reconsideration of this viewpoint is both necessary and important.

Pierre Musso's work on network ideology provides the occasion for this project. My text offers a commentary on Musso's essay and will explore in particular its contributions to the field of the philosophy of technology, around the theme of technical mediation. From this viewpoint there are several relevant aspects to Musso's work. First, it offers a historical perspective. It covers a larger time span than is common in most contemporary philosophy of technology, and the emphasis is on an early period that is otherwise largely neglected. While writings from the 1950s onwards, by Heidegger and Ellul for example, are commonly acknowledged as major points of reference, the nineteenth century is rarely mentioned. Musso's work, with a focus on the technocratic pioneer Saint-Simon (1760–1825), looks into the earlier history of thinking about technology. Secondly, the enlarged time span also lends itself to an alternative philosophical analysis. Musso's work contributes to a better understanding of what could be called the *utopian view* of technology in the early philosophy of technology and, as such, could offer a more profound and pertinent understanding of what is otherwise often referred to as the neutrality and

S. Dorrestijn (✉)
Saxion University of Applied Sciences, Deventer and Enschede, The Netherlands
e-mail: s.dorrestijn@saxion.nl

instrumentality of technology. Thirdly, Musso's work has implications with regard to contemporary questions concerning technology as well. The utopian conception may have prevailed in the early period but appears still to be widespread. In particular with respect to discourses about the Internet, Musso offers a critical assessment of its utopian aspects.

The following section first introduces some central themes from Musso's essay. In the extended middle part I will elaborate on the *utopian view of technology* from (1) the perspectives of historical technical utopias, (2) as one figure of technical mediation in the history of the philosophy of technology, and (3) with regard to actual problems and stances regarding technology.

From there I will further comment on the contribution of Musso's work to the philosophy of technology. Here, I will also outline some questions and points of critique, focusing on two major themes. The first theme concerns Musso's suggestion that it is possible to distinguish between an appropriate and desirable social utopian striving and a worn out form of technical utopianism. It is a question whether such a distinction is feasible, and indeed whether it is desirable. The second theme concerns the network. To what degree does the network indeed occupy such a central place in the ideology surrounding technology? Is the network to blame for the ideological exaggerations of techno-utopianism? Or is the network notion more ambivalent, and could it also offer a framework for critical approaches to technology?

Musso's Critique of Network Ideology

In order to discuss the contribution of Pierre Musso's work to contemporary philosophy of technology, this section will provide a short summary of his analysis of the network and the utopian ideology connected to it.[1] Musso argues that in the early eighteenth century, the social and political importance of technology and engineering began to attract philosophical and political attention, of which the work of Saint-Simon offers an outstanding example. It was the era of the building of large technical networks: the telegraph and railways. Together with the emergence of such concrete network technologies the network notion also became important as a thought concept. The network concept was soon invested with amplified meaning, namely the universal association between people and radical social improvement. Such exaggerated hopes and beliefs in the network amounted to a technical utopia. Today's discourses on the Internet, the "network society" and "cyberspace", are marked by similar and by now "worn out" utopian beliefs about network technology. Musso apparently concludes that the "social utopia" of Saint-Simon remains

---

[1]This section summarizes some of the important points of Musso's text in this volume. It seems unavoidable that this summary may seem unnecessary, or alternatively too brief, depending on whether or not readers are acquainted with Musso's essay.

desirable, even essential, but the excessive utopian expectations concerning network technologies that followed in the wake of Saint-Simon are best abandoned.

While the specific importance of Saint-Simon will be discussed later on (in the historical section on technical utopias), I will here give a concise account of Musso's analysis of the network and of the utopian ideology he sees connected to it.

The Network as Concrete Technology and as Thought-Model

In Musso's research on the impact of technology the focus is on the network character of technology. He analyses the network in a double sense, as a concrete technical phenomenon and as a thought-concept for interpreting our world. First, as a technical phenomenon, the network-structure is an important characteristic of many inventions of the modern era: the railway, telegraph, electricity, and recently the Internet. Secondly, we have also come to see and think in terms of networks. We interpret the world and ourselves in terms of network structures. The network functions as a "technique of the mind", an idea that carries and frames our thoughts.

We have not always interpreted the world in terms of networks. Musso recounts how prior to the use of the network model, the prevailing model was for a long time that of the tree. The main difference between these models is that the network conceptualization involves a grid with connections in all directions, while the tree model involves branches growing out of a central trunk. Such different conceptualizations affect ideas about the structure of reality.

In a development that started long ago, the network model has replaced the tree model. Musso's archaeology of the network model shows variants of the network model which relate to the history of technology. Beginning in Antiquity, the concept of the network was related to weaving – to threads and knots. Later, in the modern era the network became a notion for understanding living organisms; blood vessels and later the nervous system, for example. Then, finally, came the arrival of the technical inventions that we understand as network technologies: the railways, electricity, and recently ICT networks. The network as a thought model became dominant with the spread of concrete network technologies such as the railways.

Musso's argument is that the material, technical network cannot be separated from its other modality, namely as thought-model. The network functions as what he calls an "inter-world" between technology and the human body (which could be understood as a place containing both, or as a linkage between the two). The network offers images that enable a combined understanding of the human body and of technology, by explaining one by reference to the other, in both directions. For example, in Antiquity Galen compared the human brain to network-technology, namely a fisherman's net. And today the computer and the Internet are often compared with the brain and the nervous system.

The network thus understood as linkage (inter-world) of body and technology shows a development in stages, which Musso analyses as follows. At first, the network envelopes the body (woven clothes). The next stage involves the network making up the body (vessels and nerves). In the final stage the network spreads beyond the body. In the form of the railways and electricity grids, the network envelopes the territory of the whole earth, all of nature. The network now transports the body over the earth and connects it with, ultimately, the universe.

Network Ideology

Using the network model to understand the place of human beings in the world is, as Musso's research shows, not simply an analysis of how things are, but is loaded with value. The network-concept gives "meaning and direction", and the tendency is towards a very positive belief in the benefits of technology. Musso speaks of "techno-messianism" and of "techno-utopianism".

In Musso's account this techno-utopian merger between technology as concrete phenomenon and its value or meaning first came about in the utopian, technocratic movement of Saint-Simon and his followers (in the nineteenth century). At that time the network fully acquired its modern meaning as a web covering the territory of the earth. In addition, the Saint-Simonians were the first to consider the connection between technology and social issues. Engineers began to imagine the construction of an ideal social body. They began to see their work of building railways and telegraph networks as a political undertaking.

This established a technocratic view in which engineers have the means with which to carry out the task of building technical networks and thereby promoting the inherent values of democracy, freedom and equality. The political task connected to the network grew, as the network became the symbol of "universal association". This attitude towards network technology and imaginary is characterized by Musso as a "cult of the network", or "network fetishism". Within the Saint-Simonian movement itself it was not uncommon to speak of a "new Church".

Ever since then, the network has retained these utopian connotations, as Musso illustrates with examples from politics: Lenin's insistence on the importance of the electricity network for the construction of the Soviet Union almost a century ago, and Al Gore's more recent enthusiastic commitment to the "information highway", or Hillary Clinton's exclamation in relation to the Arab Spring in 2011 that "the Internet is freedom!". Every technical innovation tends to revive the "myth" of humanity being connected and united by network technology.

Technical Mediation and the Figure of Utopian Technology

Musso's analysis of the role of the network as simultaneously a concrete technology and a thought model that allows humans to understand themselves in relation to their material environment may very well count as an instance of research into technical mediation. The term *technical mediation* is frequently used to denote the influences of technology on human existence. This usage is evident in the works of contemporary philosophers of technology such as Don Ihde (1990), Andrew Feenberg (2002), Bruno Latour (1994), and Peter-Paul Verbeek (2005). And technical mediation was also a pivotal term in the research on the cultural effects of technologies by media studies pioneer Marshall McLuhan (2003 [1964]). The notion of technical mediation refers to the phenomenon that the human way of existence and ways of living are mediated by technologies. It may also refer to a further claim, or general idea, that human existence is essentially dependent on and marked by technology.

Seen from this angle, Musso's research focuses on the nineteenth century and the network technologies emerging at that time and recounts how these had a mediating impact on human existence and culture. His work helps to show the relevance of the technical mediation concept in the history of thinking about technology, even before the term itself became popular in the philosophy of technology.

There is a tendency, notably in the work of Verbeek (2005, 2011), to conceive of technical mediation as a novel and improved theory of technology and humans that has developed with the move towards an "empirical philosophy of technology" (Achterhuis 2001). In this way Verbeek attempts to distance himself from earlier approaches, such as the referential philosophies of technology of Martin Heidegger and Jacques Ellul, which are now deemed too all-embracing and negative. At the same time the technical mediation approach intends not to fall back to the pre-philosophical, ordinary understanding of technology as mere neutral instruments, which was the outlook both Heidegger and Ellul explicitly reacted against.

It is however questionable whether it is fair and necessary to reject other (earlier) approaches to technology because of their supposedly incorrect understanding of the technically mediated character of human existence. It is not the analysis of technical mediation that is so very different. Søren Riis (2008) claims, for example, that contemporary thinker Latour and classical philosopher of technology Heidegger are very much comparable in terms of understanding the interdependency of humans and technology. And Robert Scharff (2012) argues for acknowledgement of the comparability of the human-technology complex as conceptualized by Heidegger and that described by the nineteenth century French philosopher and sociologist August Comte. The connection from Latour to Heidegger to Comte by Riis and Scharff suggests continuity in the understanding of the theme of mediation of human existence by technology.

Moreover, Scharff argues that the contemporary empirical approach may even suffer from "too much concreteness", which may hinder a critical stance. Contrary to the claim that the empirical approach *discovered* the mediation idea, one can with equal right hold that the loss of a critical concern indicates a *narrowing down* of the understanding of the mediating effects of technology.

The understanding of technical mediation is not exclusive to research of the empirical type. The most important difference between periods and schools is not whether technical mediation in itself is theoretically acknowledged, but rather bears upon the ethical evaluation that accompanies and feeds the analysis of mediation.

The classical philosophers of technology did see a deep correlation between technology and human existence (which means an analytical, theoretical acknowledgment of technical mediation). What is so distinctive of their time is that they saw so much danger in the form this correlation was taking.

And the ordinary view of technology as neutral instruments is not the essence of all thinking about technology before classical philosophy of technology. Rather, an ordinary view and a more profound philosophical understanding always coexist. In early philosophy of technology the deeper meaning was that technology is indispensable to the viability of human existence. Humans are born faulty, and technology is necessary to make them viable and complete. This view definitely does acknowledge mediation of human existence by technology, but this mediation is only somewhat obscured because it is seen to be naturally good and not in the least problematic.

The present essay is meant to contribute to a broad cultural philosophical approach to technical mediation. The perspective adopted is historical and hermeneutical. It is not opposed to, but includes earlier conceptions of the influences and meanings of technology. The difference between our favorite contemporary conceptions and earlier ones is not whether the mediated character of technology has or has not yet been discovered. Instead the question is just how people have conceived of the influences of technology. There may be a variety of conceptions reflecting the persuasions of people or the historical or cultural circumstances and dominant views. Such a hermeneutical historical approach considers which different "figures of technical mediation" have been "discovered", "configured" or "conceptualized" in thinking about and coping with technology (Dorrestijn 2012a, b).

The question is, therefore, not if technical mediation is already or not yet fully acknowledged and understood, but what different figures of technical mediation are prominent. In a very general and sweeping historical overview three figures appear. First, a utopian view of technology; where technology means the way towards perfection or completion of human existence. Secondly, a dystopian view of technology in which technology appears to accumulate into a system that threatens to take command. Thirdly, a prominent conception today is that technology is neither ultimately positive nor negative but always ambivalent. In this view the challenge is to acknowledge the hybridity of human existence and technology without dystopian despair or utopian belief.

To present these figures of technical mediation in a historical progression does make some sense, because they reflect the prevailing conception of an era. At the same time, all three figures are present together at any time. They obviously also form an ever-present dialectical triangle, where the ambivalent view combines and relieves the opposing utopian and dystopian conceptions.

Of these three general views, the utopian view of technology and the *early stage* in the philosophy of technology are underrepresented in the philosophy of technology. Read against this background, Pierre Musso's work appears as a contribution to the understanding of the utopian figure of technical mediation. Saint-Simon and his network-enthusiastic followers are important thinkers who articulated the typical understanding of how technology mediates human existence, in what could be called early philosophy of technology. Musso's work thus contributes to acknowledgement and elaboration of utopian technology as a historical stage in the philosophy of technology, and also as a living view on technology today (both in the form of explicit utopian belief and as an implicit deeper layer under the instrumental conception of technology).

History, Philosophy, and Actuality of Utopian Technology

Utopian technology as one main *figure of technical mediation* deserves more attention in the philosophy of technology. The following outline will successively pay attention to the history, philosophy and actuality of the utopian view on technology.

For this I will draw on Musso's work, but also introduce other historical and philosophical material. As we have seen, Musso extensively explored the meaning and development of the notion of the network. Utopianism, the other prominent notion in his research, is not treated as extensively. I will start with providing some background on the concept of utopia and recall a few historical examples of technical utopias (where Musso recurs because of his work on Saint-Simon). Next I will look into the philosophical analysis of technology, which underlies or accompanies the techno-utopian projects (this provides a context for Musso's analysis of the network). Lastly, the actuality of tech-utopian thinking is (concisely) discussed with reference to Musso's critique of the network society and cyberspace.

Technical Utopias in History[2]

The term utopia was introduced in the book *Utopia* by Thomas More from 1516, the theme of which revolved around a new society built on an Island. The book started a genre that has produced many novels (from *New Atlantis* by Francis Bacon to *The possibility of an Island* by Michel Houellebecq) and movies (*Blade Runner, The Matrix*) (there is clearly an overlap with *science fiction*). Thomas More coined the term utopia himself. The scholar of utopias Thierry Paquot (2007) explains that More used a pseudo-Greek rendering of the Latin term *Nusquama* (which was the title of an earlier draft and means not existing place or land, as nusquam means nowhere/on no occasion). More's construction "utopia" was meant to refer to both *ou-topos* and *eu-topos,* giving it a double meaning. Utopia is simultaneously "a land that doesn't exist on any map (*outopia*), and would be the best in the world (*eutopia*)" (Paquot 2007: 6).

---

[2]This section on technical utopias uses parts from the chapter "The Legacy of Utopian Design" of my PhD thesis (Dorrestijn 2012a).

Claeys and Sargent (1999), editors of *The Utopia Reader*, emphasize that utopias are characterized by "human contrivance". This distinguishes utopias as a modern genre that is distinct from "myths", dreams of "Arcadias" and "earthly paradises", as well as from the temporal transformation of society during "festivals". Utopian thinking thus refers to the tradition that started with Thomas More, which concerns not dreams of "sensual gratification", but is about a radically different society, "humanly contrived" and intended to be realized (ibid.: 2–3). Moreover, it is this activist utopian tradition that is very relevant for the philosophy of technology, because it is in this tradition of utopian thinking that technology often plays an important role.

Utopias where technology is explicitly applied as a means for contriving a new society can be called "technical utopias". A form of utopian thinking about technology and government can be traced as far back as Plato in ancient Greece. However, Francis Bacon's *New Atlantis* from 1627 is the archetypal modern technical utopia. Two more examples I will mention are the utopian plans and strivings of Jeremy Bentham and of Henri de Saint-Simon, both active in the decades around the turn of the nineteenth century. Saint-Simon will also lead us back to Musso, whose work on Saint-Simon will be a main reference.

Bacon's New Atlantis

New Atlantis (1627) is a posthumously published book by English philosopher, statesman and natural scientist Francis Bacon (1561–1626), who remains famous for being an early advocate of the modern, empirical method of scientific research at the time of the Scientific Revolution. The book, which has become the archetypal *technical utopia*, tells of the adventures of the crew-members of a ship who after a storm at sea find shelter on an island that apparently sustains a very advanced society. The novel starts with a long description of how the shipwrecked visitors meet with the islanders, a virtuous people with a pious Christian faith.

Ultimately the visitors to New Atlantis hear everything about Salomon's House, the state agency for scientific and technical research and state government. The technical inventions conceived in Salomon's House include: food conservation caves, industrial production of foods and beverages, health conservation and life prolongation centers, breeding of modified species, light from new sources in all possible colors, distant seeing devices, artificially produced materials, instruments that produce artificial sound and music, etc. (Bacon 1999).

Readers of today are often impressed to see how accurate many of Bacon's forecasts have proven to be. Our world, in many ways, resembles the utopia of *New Atlantis* (Lintsen 2002). It seems as if Bacon's plans have actually played a guiding role in the construction of the modern industrial world.

Benthamism and the Panopticon

A second influential technical utopia is the Panopticon plan (from around 1791), created by the English jurist, philosopher and social reformer Jeremy Bentham (1748–1832).[3] The Panopticon is a circular building, which allows for continuous inspection (the combination "pan-opticon" alludes to "all-seeing"). The Panopticon, which Bentham's brother, an architect, helped design (Bentham 1843, IV: 40), consists of cells in a circle, six floors high, built around a central watch-tower. On the inside, directed towards the lodge, the cells would be largely open; only a light iron grating was planned. The central watchtower itself would be covered with a transparent curtain "that allows the gaze of the inspector to pierce into the cells, and that prevents him from being seen" (Bentham 2002: 12–13; cf. Bentham 1843, IV: 44).

This "simple architectural invention" (Bentham 1843, IV: 39; cf. Bentham 2002: 11) would make possible effective surveillance and control of people in prisons, asylums, schools and ultimately society at large. In a Panopticon people have no possibility of doing wrong, and as Bentham was convinced, this would eventually also remove the will to do wrong. "Benthamism" refers to a rationalist vision of ethics and government, based on the principle of "utility" (utilitarianism), to which the Panopticon plan is intricately connected. In the existing world people cannot always know the consequences of their actions. And actions that go against the principle of promoting happiness for the community may go unpunished or even prove beneficial in the short term for the actor. In the ideal world one would always immediately experience the right consequences of one's deeds. The Panopticon design shapes such a world where everything and everyone is always visible. In that ideal world people will always act rationally, in accordance with the rational moral principle of utility (maximizing happiness and preventing pain).

---

[3]Bentham began to write about the Panopticon in a series of letters during a stay in Russia in the year 1787. A book edition of these letters appeared in 1791. Later the texts were republished together with extensive "postscripts" in Bentham's collected works (Bentham 1843, IV). A concise edition of the Panopticon appeared in 1791 in French (Bentham 2002). This French text was an abbreviated version of the English manuscript, including some ideas from the postscripts, edited by Étienne Dumont, a friend of Bentham's. It was prepared for the French National Assembly (set up after the French Revolution of 1789). Cf. Bentham (1995) for a contemporary English edition of a selection of these texts.

Bentham's plans are not written as a novel, but are presented in letters and reports including detailed technical drawings directed at prison owners and national governors. These were clearly created in the hope that they might be realized. Although it is a matter of debate to what degree Benthamism has become a reality, it is certain that this way of thinking has been influential. There are many examples of dome prisons inspired by the Panopticon. Moreover, the ideal of ubiquitous inspection has spread in our societies in new technical forms like surveillance cameras. In addition, citizens around the world now all walk about with smart phones with cameras, which can be used to record any incident the smart phone carrier may encounter. One could say that all the smart phone carriers are in a position to inspect each other, and are thus in accordance with Bentham's ultimate ideal, namely ubiquitous mutual surveillance of people without guards.

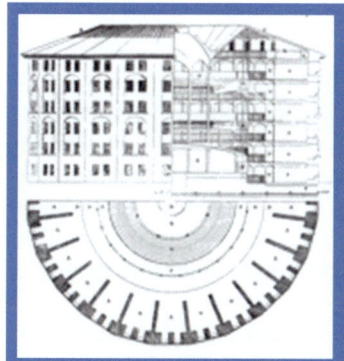

Panopticon plan and prison built after Panopticon model on Isla de la Juventud, Cuba 1926

Saint-Simonism

Definitely influenced by Bentham, and displaying a similar activist attitude of reform, is the third utopian thinker to be addressed here: Claude-Henri de Rouvroy, Comte de Saint-Simon (1760–1825). For an overview of his work and life, Pierre Musso's scholarship is significant, for example his monograph *Saint-Simon, l'industrialisme contre l'État* (2010). Henri de Saint-Simon (as he is usually referred to) was born into an aristocratic Parisian family and became a philanthropic socialist thinker and political publicist.

Today Saint-Simon is considered a pioneer of technocratic government. He proposed a reorganization of the state according to the principles of industry. He believed very much in the benefits of science and technology, properly employed, and called for "industrialism": industry delivering the principle for the construction and government of society at large. After his efforts to convince the leading liberal political movement of his ideas about the "industrialization of politics" failed, he radicalized his ideas and strived for a "politicization of industry" (Musso 2010: 127): engineers were to enter politics, and the state was to be transformed in line with the factory model, to be managed according to principles of efficiency and economic profit (ibid.: 106). Saint-Simon proposed a reorganization of parliament with the establishment of three chambers: a "Chamber of Inventions" consisting of engineers who would design public works, complemented with chambers for control and execution (ibid.: 138).

His musings on an industrial society involved not only production and economy but spiritual and religious life as well. Science was hailed by Saint-Simon as a new religion, the successor of traditional religion. This theme returns in the work of Auguste Comte. He saw the emergence of a positivistic scientific stage in civilization that would replace the previous metaphysical stage, which likewise had displaced a religious stage (Comte, early in his career, served as Saint-Simon's secretary). However, Saint-Simon himself chose not to abandon religion, for later in life he called for a "new christianism". In so doing, he tried to associate himself with traditional religion, in the hope of bringing about a revolution from the inside. On the one hand this shows that for Saint-Simon technocratic government was connected to ethics, and not just concerned with cold economic computation. Indeed, Saint-Simon's project was driven by very strong social and humane, philanthropic values (ibid.: 151). On the other hand, it also shows again Saint-Simon's radical, utopian aspirations: after calling for radical transformation of the state he embarked on what could be considered the even more ambitious project of revolutionizing religion.

Saint-Simon is widely recognized for his utopian ideals, and his work is an emblematic example of utopian social engineering. The writings of Saint-Simon influenced Karl Marx and Friedrich Engels as well as Auguste Comte. This makes Saint-Simon a pioneer of sociology, the then new science of social relations and the arrangement of society. His views were socialist and technocratic. He asserted that society should be reorganized following the principles of technical design and production. The communist principle of work according to capacity, reward and need also stems from Saint-Simon. He had a large following of *Saint-Simoniens*, who after his death endeavored to "Saint-Simonize" France (Paquot 2007: 41–42). Musso's essay on the network ideology illustrates just that, by recalling the Saint-Simonians' commitments to building technical networks and spreading the ideals of universal association.

These are three characteristic examples of technical utopias from the modern era. Utopian thinking has lived on through the twentieth century and into our time. In literature and film the utopian genre has changed, inasmuch as utopian dreams have been compromised by the danger of a reversal into dystopia. Famous examples include Aldous Huxley's *Brave New World* and George Orwell's *1984*. Outside literature, real-world examples of twentieth-century utopianism include the utopian city planning of architects such as Le Corbusier. Another more recent example is the *transhumanist* movement, the members of which believe that by technically enhancing ourselves we will evolve into a cyborg kind of being, no longer human but post-human. And of course the topic of Musso's work, the networked world of the digital age, also nourishes utopian views.

Early Philosophy of Technology: The Utopian Figure of Technical Mediation

Having looked at historical examples of technical utopias, I will now elaborate on the conception of technical mediation in this utopian view. The philosophy of technology has increasingly become an acknowledged branch of philosophy, starting with the classical works of Heidegger, Ellul, Mumford and the like. Prior to the work published by these scholars, the philosophy of technology was marginal and not a main topic for most well-known philosophers. It is, however, very useful to look at the few examples that did appear and explore how this early philosophical research on technology connects to the cultural meaning of technology that we have seen expressed in technical utopias. This section will provide further elaboration of a theme that was already to the fore in Musso's analysis of the network, namely technology as extension of the human body and simultaneously as enabler of self-understanding. And we will see how this can be understood in terms of both an analysis and evaluation of technical mediation.

For a concise discussion of the early philosophy of technology I will refer to Ernst Kapp and briefly to Karl Marx. This is not an unusual choice, as these philosophers recur frequently in historical overviews (Mitcham 1994; Ihde 2009). Somewhat different is the focus on technical mediation. Verbeek, in his version of a philosophy of technical mediation, claims that recognition of technical mediation is an achievement of the contemporary empirical approach. He positions his approach against classical philosophy of technology, as well as against the pre-philosophical idea of neutral technology (Verbeek 2005). When he does mention Kapp and the early philosophy of technology (Verbeek 2012), he adds Kapp to the list of approaches unable to grasp mediation because Kapp's work is based on dialectic philosophy.

From my standpoint dialectic philosophy is not a stumbling block. To me it seems that the dialectic philosophy of Hegel, especially as used *upside down* by philosophers like Marx and Kapp, could be considered a prototypical philosophy of mediation. The early philosophy of technology may be rather abstract and overly optimistic – but there is little question that it is a philosophy of technical mediation. Any cultural historical and hermeneutical approach to technical mediation should consider the different ways in which the mediation effects of technology have been grasped and expressed. So, the question is what was the prevailing *figure of technical mediation* in the early philosophy of technology?

Ernst Kapp's *Grundlinien einer Philosophie der Technik* of 1877 is commonly referred to as the first study explicitly titled a philosophy of technology (Kapp 2007 [1877]). Kapp's most famous insight is that every technology can be considered an "organ projection", an exteriorization of some function already found in the human body. A hammer is the extension of the human fist, a wheel of the legs, the telegraph of the nervous system. Kapp systematized this theme of technologies as extensions of human functions, even if the idea's roots can be traced to Plato, and Samuel Butler's *Erewhon* (see Chamayou 2007).

Although the details and scope of Kapp's claims have always been met with severe doubts, the analysis of technologies as extensions of man and the question of social and ethical consequences has remained a classic theme. It recurred in a variety of approaches to thinking about technology: in philosophical anthropology (Gehlen 1980), media studies (McLuhan 2003 [1964]), futurology (Kelly 2010), and philosophy of mind (Clark 2008). Next to the *functional analysis* from the outside perspective there is also a tradition of *experiential analysis* of technical extensions from the user perspective. Here I am thinking of the research focusing on the *embodiment* of technologies and *body techniques*, from the classic work in anthropology by Mauss (2006 [1936]), and in phenomenology by Heidegger (1996 [1927]) and Merleau-Ponty (1962 [1945]), to contemporary elaborations and applications (Ihde 1990; Warnier 2001; Tenner 2003; Kockelkoren 2003; Dant 2005; Noland 2009).

The analysis of technologies as extensions of humans by and after Kapp is relevant in the context of the mediation of human existence, because the technical extensions are conceptualized as indispensable parts of human existence. If technologies are in the ordinary view neutral, innocent add-ons that we deliberately design and use, than surely Kapp's analysis cannot be reduced to this. The distinctive point about Kapp is precisely that he sees technology as an indispensable mediator for the constitution of human existence, whereas we ourselves do not altogether see through or control this.

The technical mediation character of Kapp's philosophy of technology becomes clear if one considers its dialectal scheme. Following Kapp, humans, in the act of producing artefacts, imitate, unconsciously in the first instance, their own organic functions. This is for Kapp the first step of a dialectic development. The second step is that after imitating their own organs in technological mechanisms, humans begin to think of themselves in terms of these mechanisms. The heart was understood as a pump, for example. And today the brain is understood as a computer. We cannot think about ourselves otherwise than by using concepts and insights gained from technical inventions. Technologies are not simply innocent, neutral enhancements of the materialistic conditions of human life, but technology is essential for the emergence of consciousness. In addition, technologies are not, initially, invented and employed deliberately by humans, but it is technology that constitutes human self-understanding in the first place.

The reason why the deeper grasp of technical mediation in the early philosophy of technology might be overlooked is the absence of a critical ethical concern about the effects of technology. Kapp was clearly more concerned with what technology is and how it develops than with ethical evaluations of its effects. Or rather, in its essence technical progress was imagined as essentially good. This seems characteristic of the time.

Karl Marx, definitely also a relevant thinker in the context of an early philosophy of technology, adopted plainly a more critical perspective towards the industrialization process of his time. But for Marx too, technology itself is not problematic. The problem was the inequality between the class of capital owners who fare well and the mass of working class people who only suffer from bad working and living conditions without profiting from industrial products themselves. The ethical concern is about fair distribution, making everybody share in the wonders of technology.

Therefore, in the early philosophy of technology the prevailing conception of the effects of technology combined with the accompanying ethical concern can be summarized as follows: *technology is the miraculous means for human completion; we only need for scarcity and unequal distribution to be overcome*. This can be called the dominant figure of technical mediation in the early philosophy of technology.

Following this insight into the early philosophy of technology it is possible to look back and see variants of the general mediation figure in the historical technical utopia projects. In Bacon's New Atlantis technology is presented as the miraculous means by which humans conquer more and more of their faults and have perfect lives. Bentham offers a variant which focuses on technology as a means for bringing human morality to perfection. Technology helps to illuminate the right associations between actions and their consequences, which may be flawed in real society because of the long chains of consequences and the difficulty of overseeing interactions between all individuals. Typical of the techno-utopian discourse, technology itself hardly becomes an ethical issue. It is considered a helpful and even essential support for the correct functioning of morality. The Panopticon will correct the morals of those imprisoned, and the same principle could likewise perfect morals in society at large. For Saint-Simon and his followers technology in the age of the network fosters solidarity between all inhabitants of the earth. The network with its inherent social and moral qualities is conceived of as a "lever for political action", as Musso writes in his essay (Chap. 2).

In the technical utopias of the modern era the discovery of this dependence of human development on technical progress was met with enthusiasm. The challenge became to take up and speed up this development as a conscious process. Technologies promise a society where there is an abundance of material goods, where the suffering caused by illness and pain may be alleviated, and where crime no longer exists. A perfect human being and a perfect society are not given from the beginning, but the promise is that they will be gradually realized. While human beings are as yet incomplete, their fulfillment, completion, or perfection lies ahead in the future as a cultural and technical project.

Characteristic of technical utopias is the unquestioned commitment to the advancement of this project of progress. This may, by the way, suggest a view that technology could always be consciously applied and controlled. That view reminds us of the ordinary, *instrumental* conception of technology (which neglects mediating effects). But again, particular to this utopian view is that the mediating effects of technology can remain unnoticed because they converge with common-sense idea of instrumental use. The use of technology is innocent, or always good, because it is in the nature of technology that it brings progress to humanity. The function of human consciousness is to accommodate this natural process of progress rather than that it should claim to be the intentional actor. In this respect there is a difference between Kapp's philosophical analysis of technology and activist utopian plans, for Kapp explicitly debunks the assumption of awareness and control. Kapp thinks that technologies are always the projection of a function of the human body: that is the nature of technology; it could not be otherwise. But this does not mean that the human inventors are aware of this. And also they are not fully aware of the transformation of themselves that is the result of their inventions.

Recognition of technical mediation is not an achievement of the newest current in the philosophy of technology. From the beginning of the philosophy of technology, the phenomenon of technical mediation has been grasped, and this grasp has never been lost. In every period there is a more ordinary, but also a more profound philosophical understanding of technology which properly acknowledges the radical constitutive role of technology for human existence. Now the technical mediation figure of early philosophy of technology has been characterized, how does it compare to later views? And what is the continuing relevance of the utopian view?

Utopian, Dystopian, Ambivalent

The utopian conception of technology has long ceased to be the prevailing mediation figure. By mid-twentieth century critical ethical concern over technology took precedence. The conception that came to the fore was that instead of being in itself good, technology rather turned out to be a great danger. The nuclear bomb is the emblematic example: a human invention but so dangerous that it can annihilate humanity. The urgency of this experience prompted many prominent philosophers to start to analyze the dangers inherent in technology. Philosophy of technology was no longer a marginal field. In this period of classic philosophy of technology the dominant conception of technology turned dystopian, as compared to the utopian view of the earlier period. Summarized in one sentence, the dystopian figure of technical mediation in classical philosophy of technology is that *technology threatens to accumulate into a system that takes control and the ethical task is therefore to set limits.*

The change of conception of technical mediation, both as analysis of and as ethical concern with the impact of technology can be illustrated by looking at the interpretations of Kapp's analysis of technology as organ projection. I follow Georges Canguilhem (1965: 127) who analyzes that for Kapp a progressive technical extension of humans is only natural, and a critical ethical questioning of technology has not yet begun. However, a recent commentator on Kapp, Benoît Timmermans, thinks there is an ethical message to be derived from Kapp's work, namely that "everything has to be done to prevent technological projection (...) leading to alienation, mechanical dependency, or resistless subjection to what we have produced, but what has become irretrievably foreign to us" (Timmermans 2003: 105).[4] Indeed, this is an eloquent expression of a fear that we recognize today. It is the ethical concern that emerged in the era of the classical philosophy of technology. But I think it does not actually represent the thought of Kapp and his time. In the early philosophy of technology the ethical concern only mentioned technology indirectly, namely from the economic perspective of scarcity and distribution. In itself the accumulation of technology was welcomed. This changed with the reversion to a dystopian view of technology. Technology, or modern technology, is accumulating into a large system that gets out of control and threatens human values.

Today neither the dystopian nor the utopian view is the dominant figure of technical mediation in the philosophy of technology. In the last few decennia there has been an "empirical turn" (Achterhuis 2001) towards a more practice and application- oriented style of research. Technology does not have a single essence given once and for all (Ihde 1990), but always holds both positive and negative possibilities at the same time. The way in which technology is implemented does matter, and it is possible to correct for its negative effects. The meaning of technology is now seen as always *ambivalent* (Borgmann 1984; Feenberg 2002; Lemmens forthcoming). In the conception of technology that is now gaining prominence the fundamental mediation of human existence by technology is acknowledged, but is evaluated in ambivalent terms, beyond utopian hopes and dystopian fears. Insight into dependency on technology is now interpreted as a call to ensure the quality of fusions and interactions with technology in a pragmatic way. Expressed in one sentence: *human existence is ambivalently mediated by technology, which prompts us to take care of our hybrid existence.*

Interpreted in this way, technical mediation is not an approach that replaces earlier more essentialist analyses of technology. Technical mediation is an enduring theme that has always been grasped, but within this theme there is a wide variety of conceptualizations and evaluations of the effects of technology. In the most general terms there is a historical development from early, to classical, to empirical philosophy of technology, where the prevailing figure of technical mediation went from utopian to dystopian to ambivalent (Dorrestijn 2012a, forthcoming).

But while there may be differences in dominance, all three figures do also coexist at any given time, forming a dialectical triangle. This is one more way of seeing how within the current conception of the impact of technology as ambivalent, the utopian and dystopian views of technology can still be acknowledged and valued. In a historical and hermeneutical approach to technical mediation, one need not denounce earlier approaches in favor of a newer and better theoretical grasp of technical mediation. Instead, the tension between the divergent conceptions and ethical evaluations of technical mediation makes the current ambivalent view all the more interesting.

---

[4]My translation.

Actuality of Utopian Technology: Cyberspace and Network Society

So far we have looked at the history and the philosophy of the utopian view of technology. Pierre Musso's work on the utopian projects of Saint-Simon and his followers helped in both respects: remembering the history as well as philosophical explication. Now I will use Musso's work to show, briefly, how the utopian figure of technical mediation remains important today. Musso discusses utopian ideas on the network at work in our age of information networks. He sees two main figures in today's network utopia: cyberspace and the network society.

Cyberspace refers to the hybridization of man and machine to the point that a form of life emerges that is disembodied (in the sense of a biological body), yet re-embodied in a liquid, electronic, information network-space. This implies a distributed-networked body as well as a networked, interconnected, distributed intelligence. The network society as proclaimed by Castells denotes a view that reminds Musso of both Marx and McLuhan and in which the Internet forms a new material basis, namely an Internet Galaxy that creates a new society and a new economy.

Musso is critical of the meaning given to technology in the network utopias of Cyberspace and the network society. He thinks that more than ever before technology provokes thought models. The idea of ubiquitous connectedness becomes a spiritual belief. The networks of the industrial age had a physical bodily connotation, and represented the possibility of universal contact and displacement. While techno-utopianism always had a religious and moral meaning, the network in our digital era has acquired an even stronger spiritual connotation. This seems to imply for Musso a leap into a fantasy world at the cost of a sense of the concrete. Musso fears that a belief in technical utopias relieves people of the burden of commitment to social and political action.

Indeed technical utopias are not of the past. Cyberspace and the network society, with their focus on network technology, may be complemented with other technical utopias. One example is the aforementioned transhumanism movement, which is committed to technically enhancing human existence to the point of accomplishing a new species beyond what used to be the human being. Space travel is another domain that fires the utopian imagination. The view on the earth from space has proven to mediate the growth of ecological consciousness, but equally the forging of utopian scale plans of geo-engineering (Grevsmühl 2014). Leaving the earth behind is of course another utopian possibility. *Mars One* is a project that recruits pioneers who are prepared to embark on a one-way voyage to Mars and start a new settlement there to relieve planet earth of its overpopulation.[5] In connection with Musso's critique above, it seems difficult to decide whether leaving the earth for Mars means the highest commitment to the cause of humanity or rather a literal

---
[5]Available at: http://www.mars-one.com/

example of escapism into a utopian belief – a way for participants to run away from the concrete challenges they face on earth.

In the remainder of the chapter I will raise two questions, namely the way in which Musso seems to preserve a desirable form of utopianism against the exaggeration of techno-utopianism, and how Musso identifies the network character of technology as the factor leading to the trap of techno-utopianism.

Save the Good from the Bad Utopia?

One important aspect of Musso's work is that it helps to bring to the fore a historical perspective and the utopian figure of technical mediation in the philosophy of technology. In the final two sections I discuss Musso's own stance regarding utopian thinking. He is very critical of the utopian connotations of technology, especially network technology. But Musso's solution is not to say goodbye to utopia. It seems rather that he wishes to restore a positive form of social utopian thinking and distinguish it from the exaggerated form of techno-utopianism. In the present section I will discuss if such a demarcation between good and bad forms of utopia is feasible. This allows us to see what an ambivalent position, somewhere in the tension between utopian and dystopian, looks like in the case of Musso, and to situate his position among current philosophers of technology.

Good and Bad Utopias?

According to Musso the point that demarcates the good and the bad forms of utopianism lies between Saint-Simon and what came after. It seems that he credits the Saint-Simonians for "theorizing the Industrial Revolution", and to a large extent also for attempting to "socialize the major technical networks". But when it comes to the identification of the technical network with an ideal organism and social organization Musso appears quite critical. Musso is not altogether explicit, however it seems that he alleges that the network ideology (or *retiology*) ended up becoming prevalent, even though no concrete analysis was made to ascertain whether the actual transformations that network technologies bring are good, or bad. The social semi-utopia of Saint-Simon (as discussed by Musso) was turned into a full-blown technical utopia by the later Saint-Simonians (right up to Proudhon and Kropotkin). They turned the concrete network technology into a "technology of the mind", conveying the myth of radical social and political transformation through the technical network. This techno-messianistic or utopian network ideology has prevailed ever since, from Lenin to Al Gore and Hillary Clinton.

This raises interesting questions in the context of the ambivalent view of technical mediation and Musso's stance. Musso contests the exaggerated form of techno-utopianism, and is aware of the antagonism of the utopian and dystopian views. He does mention that the dystopian and the utopian views are akin, that utopia can reverse into dystopia, although he does not elaborate on this idea. Now the question is if the way out of this position is to give up utopian thinking altogether, or if restoration of a moderate utopian inspiration is more desirable. Is it possible to save a moderate utopian vision from the worn out utopianism that becomes antagonistic and deceives rather than inspires concrete social and political commitment? Also, is social engagement actually dependent on utopian inspiration, or can utopia be dismissed without a loss?

Philosopher of utopian thinking and technology, Hans Achterhuis (1998, 2001), has more explicitly than Musso elaborated the notion of the antagonism between utopian and dystopian views. Achterhuis speaks of a "utopia/dystopia syndrome". When we think or debate in terms of ideal situations the result is often an insoluble dualism of the extreme positions of utopia and dystopia. For Achterhuis this antagonism is inevitable. Any utopian plan will reveal its dystopian side on the way to realization. Partly, Achterhuis would agree with Musso that utopian thinking leads away from engagement with concrete reality. But whereas for Musso the loss of concrete social engagement is the danger of a wrong kind of utopian thinking, according to Achterhuis any form of utopianism brings the inherent risks of totalitarian repression.

This marks a difference in conceptions of utopia. Some would say that it is necessary to cherish the picture of utopia, like one says that it is important to have ideals. Even if utopia may reverse into dystopia, in itself the meaning of utopian thinking is innocent and even necessary. This is for example the position held by the scholar of utopias Thierry Paquot (2007) and seems to be implied in Musso's view. By contrast, Achterhuis' study on the legacy of utopia (Achterhuis 1998) expresses a much more suspicious and critical position. Beyond motivating people to improve societies, utopian thinking has also led to some of the crudest regimes on earth. The belief that a radically different world could be constructed, purified of crime, laziness, inequality, etc. has made people engage in forcefully and cruelly purifying societies: the totalitarian aberrations of Nazism and communism. In Achterhuis' analysis the connection between utopia and dystopia is so intricate that one cannot have utopian inspiration without dystopian danger.

The way out proposed by Achterhuis therefore differs from Musso's. Achterhuis has encouraged the *empirical turn* in the philosophy of technology, which involves the inclusion of more case studies in philosophy, or interdisciplinary collaboration with empirical sciences. In comparison, Musso's approach remains entangled in discourses about the meaning of technology. Musso calls for a moderate social utopian engagement, but in a rather theoretical, discursive way, at the cost of a more empirical orientation.

Varieties of Ambivalence

Within the ambivalent view, which characterizes contemporary philosophy of technology, some are more utopian and others more dystopian in their outlook, and some are truly divided. For example, Kevin Kelly (in a way) and Peter-Paul Verbeek tend towards the utopian. Andrew Feenberg is divided: generally suspicious in connection with the social effects of technology, but positive about the possibility of change and almost somewhat credulous in relation to the democratic potential of the Internet. Tending slightly towards the dystopian side within the ambivalent position, we find for example Bernard Stiegler and, perhaps, Pierre Musso.

In the case of Verbeek, it seems that his repeated claim that the dystopian view is outdated makes him leap to a position that through lack of unease appears to be rather optimistic, with traits of utopian thinking. To overcome the strong dystopian view of classical philosophy of technology, Verbeek's strategy is to denounce the theoretical analysis of classical philosophy of technology as inadequate. The more adequate alternative would be his theory of technical mediation. This strategy implies that Verbeek thinks that an accurate understanding of the relationship between humans and technology provides simultaneously an adequate evaluation of technology. The inadequate analysis of classical philosophy of technology led to the inadequately dystopian view of technology. *Mediation theory* would be necessary to correct this mistaken evaluation.

This is however questionable. The relationship between *analysis* and *evaluation* of technical mediation is not one-way. Theoretical apprehensions are answers to an ethical concern just as much as, the other way around, ethical evaluations derive from theoretical understandings of states of affairs. Verbeek's strategy of redirecting the debate over technology from evaluation to analysis drives him to the conclusion that ethical concern about technology is only valid if it follows from analysis. To stand up against technical developments is a sign of inadequate analysis.

The mirror image seems also to apply for Verbeek: an adequate theoretical understanding of technology appears to offer *ethical reassurance*. When we know better how technology shapes our existence, we may also more easily accept these processes. Verbeek's attitude of optimism and reassurance is a trait that belongs to utopian thinking. Reassurance that we are part of a natural development of progress belongs to utopian thinking, while dystopian thinking is characterized by the shocking discovery that progress looks evil. Undoubtedly this is one reason for the fact that while Verbeek claims an ambivalent position that seeks the right *balance* (Verbeek 2014), this is not always recognizable to his readers. The effect, perhaps unintended, of Verbeek's all too strong contestation of the dystopian view of technology is that his version of the ambivalent position tends rather to the utopian side.

A comparable stance is that adopted by Kevin Kelly. In his essay *What Technology Wants,* Kelly (2010) gives a name to the process of technical evolution that means a transformation of human existence: The Technium. Kelly differs from Verbeek in that he starts from the viewpoint of technical enhancement of human existence, reminiscent of the utopian conception of technology. But interestingly, in the course of his analysis he brings in a remarkable degree of ambivalence. While he remains very much aware of the many difficulties and dangers, he nonetheless inclines towards a view that technology means more improvement than danger, but only by a slight margin.

As Musso mentions, the Internet is a technology that has given rise to exaggerated expectations, for example in Castells' work on the network society. A trace of the same kind of utopian connotation to the Internet can also be found in the philosophy of technology of Andrew Feenberg and Bernard Stiegler. These two thinkers may count as examples of truly ambivalent scholars, not to say that they are divided. Their hope for a more democratic future of technical development is the counterpoint to a fairly critical judgment of the present situation. Stiegler (2010) in particular is worried and negative about the transformation behind our backs of our way of living, of which we are often unaware and which we cannot control, because so much power resides with large commercial enterprises.

So, what does the ambivalent position of Musso look like? It seems that Musso's version of ambivalence is also one of great tension. Although there is clearly a lingering utopian view, overall his view seems to tend towards the dystopian side. One valuable aspect of this position is that it makes clear that the ambivalent position can include divergent conceptions and evaluations of the effects of technology on human culture. This means an approach to technical mediation that is historical and hermeneutical and acknowledges different views. Musso however remains strongly entangled in the discourse on technology. His work does not offer a concrete insights into alternatives for practical change. What he does allude to is the restoration of a social utopianism, but as I have discussed, the question remains as to whether that removes the danger of utopian thinking.

Is the Network to Blame?

The remark that Musso's approach becomes entangled in discursive analysis brings us back, in this last section, to the network. Musso places enormous weight on the network. It is because of network technology that a techno-utopianism emerged. He asserts that the network has become a technology of the mind. The network functions as a metaphor in thinking. While perceiving and considering the world around us we use the network as a model for how our sight is directed and how we might interpret what we see. We see connections everywhere and interpret these as progress towards universal association, transparency and freedom, while at the same time it also raises the suspicious counter-view of ubiquitous control and domination. Is the emphasis that Musso puts on the network pertinent? Is the network indeed the determinant par excellence, or is it less important, and just one point of access to utopian thinking about technology? Is it possible to avoid the trap of the network?

It appears that the relevance of the network as a metaphor is at the same time more and less than in Musso's account. It may be less, for it is questionable if the network needs to be emphasized as the technological development that is the cause of utopian worship of technology. But more remarkably, it may also be even more relevant, because the network does in a way also haunt Musso himself and may explain the entrapment in discourse at the expense of practical relevance. To elaborate on this I will compare Musso's network critique with Bruno Latour's network thinking.

Networks Reign Everywhere

The suggestion in Musso's work is that avoiding the trap of the network with its connotations would save us from a leap into techno-utopianism. Is it possible to avoid the trap of the network? And is it necessary? One reason why it seems impossible is that the network may have infiltrated our world even more deeply than is suggested in Musso's analyses. The network is indeed a model that shapes our perceptions and thoughts, as Musso has shown. This is so true, that it is even part and parcel of Musso's own method of analysis. Musso's highly informative account of the network metaphor in discourse about technology could be called a discourse analysis that traces the crossings and connections of the network concept. As an approach and research method it is therefore itself also dependent on network-thinking. This only reinforces the view that the network is indeed all-important, but it makes the call to avoid network-thinking implausible.

But does the omnipresence and unavoidability of the notion of the network bring with it all the dangers of techno-utopianism? Does it imply that one gets trapped in discourse and the utopia/dystopia syndrome? This is perhaps not true. It is remarkable that one of the most prominent thinkers about technology, Bruno Latour, actually combines a conceptual framework that draws heavily on the network concept with a research orientation towards practice and away from any grand claims as in the utopia/dystopia scheme. Since the 1980s Latour has been a prominent proponent of a research approach called Actor-Network Theory. The approach has become so well known that the term network will remind many researchers of technology and culture of the Actor-Network Theory. In a monograph on Latour by Graham Harman (2009), Latour was given the title "Prince of networks".

The actor-network method seeks to overcome any presumptions of a fundamental difference between humans and non-humans, things. Humans should not be seen as subjects or actors in control. And things should not be seen as passive objects, but their role in the constitution of an event or course of action should be recognized. Humans and non-humans are considered both alike as actors (or "actants", in terms inspired by semiotics), namely as contributors to action. This has resulted in research approaches that involve mapping actor-networks, where Latour's advice was to "follow the actor" and to engage in detailed historical and ethnographic research (Latour 1987). The result was often a fresh look at the role of things and materiality in society, politics and morality – a role overlooked or dismissed by the established approaches in the disciplines of sociology, political science, and philosophy.

Latour has given the network a connotation quite different from that in Musso's analysis. Whereas Musso shows how examples of network technologies led to overly general thought concepts, for Latour the network notion helped to find a path back in the other direction: from the all too general discourses towards empirical research into the details of the associations between concrete technologies and humans. This is also how Hans Achterhuis was inspired by Latour. Achterhuis' call for an empirical philosophy of technology as a way of overcoming the utopia/dystopia syndrome was strongly influenced by the orientation towards case studies and the deconstruction of thought-concepts in Latour's Actor-Network Theory.

Value Pluralism and Diplomacy

Recent developments in Latour's work may provide further insight into how one may adopt a critical position and a form of social engagement even when one is immersed in the network. For a long time Latour advanced Actor-Network Theory explicitly and in sweeping terms as a contestation of the modern world picture in philosophy and social sciences. In his recent work *An inquiry into modes of existence,* Latour (2013) follows up his contestation of the prevailing modern self-understanding with a "positive" description of what he thinks would be an alternative, better understanding of ourselves as moderns. This extensive book offers a broad philosophical interpretation, synthesis and further elaboration of his earlier research approach. Latour's project is of interest for us here as well, because he explicitly gives a twist to his earlier work.

Overall the book develops a non-essentialist understanding of being. The enquiry is about describing and comparing different "modes of existence", regimes of enunciation of what is. The provisional list of modes of being contains fifteen more or less distinct patterns in the way things and events obtain a certain degree of subsistence, even if these beings evolve and their existence always remains fragile.

Latour's research was often understood as merely opposed to values – this central concern of the human sciences – even if Latour himself always wanted to contest the very distinction between facts and values. In introducing the modes of existence Latour remarks that his case studies which mapped actor-networks often met with controversy and contestation. The controversy was never so much about his findings on the level of empirical and historical description, but rather it appeared that actor-network descriptions always engendered clashes over values. The actor-network description thus appeared to serve as a detour that releases people and makes them aware of their stance. Critical philosophy and anthropology can use this detour by the network analysis to raise awareness of conflicting values.

The role of critical philosophy today, according to Latour, is to perform an anthropology of ourselves as moderns. This means to speak eloquently about matters of concern with the people to whom those concerns matter. Speaking eloquently means, when successful, that mere conflicts and confusion are surpassed and become acknowledged as value conflicts in a pluralistic world. In the past actor-network theory took a certain pride in confusing the self-understanding of the moderns, although it might indirectly also have provided inspiration for practical engagement. Latour now explicitly engages on the path of "diplomacy": mediating and finding compromises in the situation of conflicting values. And what is more, insight into the network character of reality plays a positive role here. This shows once more that the network may not be to blame for the loss of social engagement and entanglement in utopian and dystopian thinking.

Conclusion

The contribution of Pierre Musso's work to the philosophy of technology is that it helps to promote an historical perspective in the philosophy of technology and it draws attention to the utopian view of technology. In this essay I have elaborated on this utopian view with regard to historical utopian projects, the early philosophy of technology and the actuality of utopian thinking about technology.

Technology is in the utopian view the obvious and necessary way to bring human existence to completion. In itself technology is wonderful and raises no ethical questions; only inequality of access to technical progress is a problem. This identification of technology as good in itself also obscures the more profound philosophical layer in early philosophy of technology. A functional conception of technology does however not coincide with the ordinary, pre-philosophical view of technology as a neutral means. To say that is to miss the associated deeper utopian meaning of technology as the indispensable mediator in the completion of human existence.

In this way I have also promoted a cultural historical and hermeneutical approach to technical mediation. Technical mediation is not one particular theoretical grasp of the relationship between humans and technology but a *theme*. If one wishes to speak of a *theory* of technical mediation, than such a theory should consist of a repertoire of different figures showing the impact of technology (analyses and evaluations), in which different historical views on technology are acknowledged. This implies that the ambivalent position in thinking about how technology mediates human society does not simply follow up earlier and less correct understandings of technology, but acknowledges and retains these as an internal tension.

While Musso's work can thus be seen as a contribution to and inspiration for current work in the philosophy of technology, there are points of critique. His work remains rather abstract, and questions arise as soon as one turns to possible practical implications. His suggestion of demarcating a worn out utopianism from a moderate inspirational form can be questioned, as can his identification of techno-utopianism with the network.

References

Achterhuis, H. 1998. *De erfenis van de utopie* [The legacy of utopia]. Amsterdam: Ambo.
———. 2001. Introduction: American philosophers of technology. In *American philosophy of technology: The empirical turn*, ed. H. Achterhuis, 1–9. Bloomington: Indiana University Press.
Bacon, F. 1999. New Atlantis. In *The utopia reader*, ed. G. Claeys, and L.T. Sargent, 118–125. New York: New York University Press.
Bentham, J. 1843. *The Works of Jeremy Bentham,* ed. J. Bowrin, vol. 4 (11 vols.). Edinburgh: William Tait.
———. 1995. *The Panopticon writings*, ed. M. Bozovic. London/New York: Verso.
———. 2002. *Panoptique: Mémoire sur un nouveau principe pour construire des maisons d'inspection, et nommément des maisons de force*. Paris: Mille et Une Nuits.
Borgmann, A. 1984. *Technology and the character of contemporary life: A philosophical inquiry.* Chicago: University of Chicago Press.
Canguilhem, G. 1965. Machine et organisme. In *La connaissance de la vie*, 2nd ed., 101–127. Paris: Vrin.
Chamayou, G. 2007. Présentation. In *Principes d'une philosophie de la technique,* E. Kapp, trans. G. Chamayou, 7–40. Paris: Vrin.
Claeys, G., and L.T. Sargent. 1999. Introduction. In *The utopia reader*, ed. G. Claeys, and L.T. Sargent, 1–5. New York: New York University Press.
Clark, A. 2008. *Supersizing the mind: Embodiment, action, and cognitive extension*. Oxford: Oxford University Press.
Dant, T. 2005. *Materiality and society*. Maidenhead: Open University Press/McGraw-Hill International.
Dorrestijn, S. 2012a. *The design of our own lives: Technical mediation and subjectivation after Foucault*. PhD dissertation. Enschede: University of Twente.
———. 2012b. Theories and figures of technical mediation. In *Design and Anthropology*, ed. J. Donovan, and W. Gunn, 219–230. Surrey/Burlington: Ashgate.
———. Forthcoming. The care of our hybrid selves: Ethics in times of technical mediation. *Foundations of Science*. [Published online first: DOI 10.1007/s10699-015-9440-0]
Feenberg, A. 2002. *Transforming technology: A critical theory revised*. New York: Oxford University Press.
Gehlen, A. 1980. *Man in the age of technology*. Trans. P. L. Berger. New York: Columbia University Press.
Grevsmühl, S. 2014. *La Terre vue d'en haut: L'invention de l'environnement global*. Paris: Seuil.
Harman, G. 2009. *Prince of networks: Bruno Latour and metaphysics*. Prahran: Re. Press.
Heidegger, M. 1996 [1927]. *Being and time: A translation of Sein und Zeit*. Trans. J. Stambaugh. Albany: SUNY Press.
Ihde, D. 1990. *Technology and the lifeworld: From garden to earth*. Bloomington: Indiana University Press.
———. 2009. Foreword. In *New waves in philosophy of technology*, ed. J.-K.B. Olsen, E. Selinger, and S. Riis, viii–xiii. Hampshire: Palgrave MacMillan.
Kapp, E. 2007. *Principes d'une philosophie de la technique*. Trans. G. Chamayou. Paris: Vrin [Translation of *Grundlinien einer Philosophie der Technik*. Braunsweig: Westermann, 1877].
Kelly, K. 2010. *What technology wants*. New York: Viking Press.

Kockelkoren, P. 2003. *Technology: Art, fairground and theatre*. Rotterdam: NAI.
Latour, B. 1987. *Science in action: How to follow scientists and engineers through society*. Cambridge, MA: Harvard University Press.
———. 1994. On technical mediation. *Common knowledge 3*(2): 29–64.
———. 2013. *An inquiry into modes of existence*. Cambridge, MA: Harvard University Press.
Lemmens, P.C. Forthcoming. Social autonomy and heteronomy in the age of ICT. The digital pharmakon and the (dis)empowerment of the general intellect. *Foundations of Science*. [Published online first: DOI 10.1007/s10699-015-9468-1]
Lintsen, H.W. 2002. Keynote lecture: Flying in the New Atlantis- and the evolution of technology. In *Around Glare*, ed. C. Vermeeren, 3–18. Dordrecht: Kluwer Academic Publishers.
Mauss, M. 2006. Techniques of the body. In *Techniques, technology, and civilisation*, ed. N. Schlanger, 77–96. New York: Durkheim Press/Berghahn Books [Translation of Les techniques du corps. *Journal de psychologie*, 32(3–4), 365–386, 1936].
McLuhan, M. 2003 [1964]. *Understanding media: The extensions of man*. Critical edition by W.T. Gordon. Corte Madera: Gingko Press.
Merleau-Ponty, M. 1962 [1945]. *Phenomenology of perception*. Trans. C. Smith. London: Routledge [Translation of *Phénoménologie de la perception*, Paris: Gallimard, 1945].
Mitcham, C. 1994. *Thinking through technology: The path between engineering and philosophy*. Chicago: University of Chicago Press.
Musso, P. 2010. *Saint-Simon, l'industrialisme contre l'État*. La Tour-d'Aigues: Éd. de l'Aube.
Noland, C. 2009. *Agency and embodiment: Performing gestures/producing culture*. Cambridge, MA: Harvard University Press.
Paquot, T. 2007. *Utopies et utopistes*. Paris: La Découverte.
Riis, S. 2008. The symmetry between Bruno Latour and Martin Heidegger: The technique of turning a police officer into a speed bump. *Social Studies of Science 38*(2): 285–301.
Scharff, R.C. 2012. Empirical technoscience studies in a comtean world: Too much concreteness? *Philosophy & Technology 25*(2): 153–177.
Stiegler, B. 2010. *Taking care of youth and the generations*. Stanford: Stanford University Press.
Tenner, E. 2003. *Our own devices: The past and future of body technology*. New York: Alfred A. Knopf.
Timmermans, B. 2003. L'influence hégélienne sur la philosophie de la technique d'Ernst Kapp. In *Les philosophes et la technique*, ed. P. Chabot and G. Hottois, 95–108. Paris: Vrin.
Verbeek, P.-P. 2005. *What things do: Philosophical reflections on technology, agency, and design*. Pennsylvania: Pennsylvania State University Press.
———. 2011. *Moralizing technology: Understanding and designing the morality of things*. Chicago/London: The University of Chicago Press.
———. 2012. Humanity in design. In *Design and anthropology*, ed. W. Gunn, and J. Donovan, 163–176. Surrey/Burlington: Ashgate.
———. 2014. *Op de vleugels van Icarus: Hoe techniek en moraal met elkaar meebewegen [On the wings of Icarus: How technology and morality develop together]*. Rotterdam: Lemniscaat.
Warnier, J.P. 2001. A praxeological approach to subjectivation in a material world. *Journal of Material Culture 6*(1): 5–24.

Chapter 6
From the Critique of the Network Symbolic Form to the Ideology of Innovation: An Appraisal of Pierre Musso's Work on the Current World Situation of Technology

Francisco Rüdiger

Networks are increasingly omnipresent. Anyone outside them, or ignorant of them, evokes surprise. We are asked to engage with them on every side, and we feel lost when they break down. Although networks, which are experienced as benign, collective and anonymous signals, may produce some anxiety, and are sometimes enveloped in mystery, there is hardly any opposition to them. Apparently, we are all in their favor, regardless of social status and political conviction. We hope they will work, and we suffer every time we cannot access them. As participants in them, we feel that monitoring their movement is not enough: we have to participate as active subjects (Vitale 2014).

In relation to the social aspect of networks (Mercklé 2011; Portugal, 2007), a growing number of us are expected to multiply our contacts and expand our connections, to react to stimuli and respond to them as promptly as possible (Parente 2004). We have been encouraged to believe that by participating in them we can bring a greater or lesser intellectual and communicative capital to our relationships, and receive in exchange the counterpart put into circulation by all (Van Dijk 1999; Castells 2013). At the same time, few people question this fluid and democratic ethos, in which persons tend to be reduced to electronic experiences, and their relationships with others to be framed in abstract and standardized protocols (Galloway and Thacker 2007).

Faced with this situation, corporations discover that they have to redefine the challenge of developing their business projects, fulfilling their mission, and serving their markets. Social resistance to innovation diminishes in an environment where change tends to become institutionalized. On the other hand, there is a growing irrationality in consumer and user attitudes, when everything is moving and

---

All translations are mine, unless otherwise indicated.

F. Rüdiger (✉)
Pontifical Catholic of Rio Grande do Sul, Porto Alegre, Brazil

Federal University of Rio Grande do Sul, Porto Alegre, Brazil
e-mail: frudiger@pucrs.br

accelerating. In a context of rampant competition, managers face the problem of our unrealistic expectations of technological innovation, and the need to reduce the economic waste involved in research and development of products and services with little or no return on investment.

For some marketing gurus, the attention which philosophy and the humanities give to understanding the imaginary may assist them in this situation. Research in these fields shows that technology and imagination, rather than being opposed to each other, may at least theoretically converge, contrary to what was assumed by former management thinking. The practical problem then, understood in terms of reciprocal shaping, is how to drive it so that their connection optimizes business action without losses to society, rather than hindering it.

Pierre Musso discusses all this with a firm hand and solid scholarship in his important work on the symbolic figure represented by the network in Western civilization and, more recently, on technological innovation in contemporary business activity. His voice is not part of the large chorus proclaiming the wisdom of networks as automatic precursors of the future. Arguments for and against tend to be kept at bay, even though the author is not completely free of them, as we will see below. The main point is to understand how almost all of us succumbed to that fiction, to become its hostages, whether in practice or in discourse, or both. Musso's conviction is that we still have the means to think about and to intervene rationally in this field.

Subjected to historical analysis and critical reflection in Musso's work, the network on the one hand takes on the shape of a social construction and a symbolic operation; and on the other hand it is also streamlined because, despite ideological exploitation, it cannot be simply discarded by thinking, given that it is open to political discussion and entrepreneurial action.

The networks that so many people talk about today are not, even in the context of computers, raw facts or data to which we may not appeal, but an objective feature of a metaphorical nature which emerged, and continues to be reworked, through many social agencies and multiple historical processes. A long-term effect of this was their conversion into an ideological motif of our epoch. A century after its modern inception, "network" has become a rhetorical expression, the axis of a *retiology* that fits well in a world experienced as continuous immersion in a flow of information and contacts subjected to perpetual expansion (Musso 2007a [2003]: 228).

All of this would be cause for despair, if we could not, in addition to applying historical and critical reflection, act rationally in the situation. Ideological operations contained in the imaginary of the network do not exhaust its meaning. Fantasies substantiating this imaginary are also a necessary part of the symbolic fictions that allow us to live in society. Research and development programs can observe signs which help them determine the rules for managing them in a useful and productive way. An example of this would be the strategic formatting of the imaginary and the specialized employment of the network in the service of technological innovation in entrepreneurial activity.

> Today innovation places become networks of interacting actors, and often large structures involving powerful poles; become networks of 'competitive' poles of excellence, which combine information and communication technologies, skills, training, organizational

quality, universities that put schools of art and marketing to interact, companies, laboratories, start-ups... (Musso 2005: no numbers or paragraphs)

Musso thinks that this renders the study of the imaginary practically relevant, since this element inhabits the development of networks. He believes it may help academics and corporations find a more rational and less ideological way to intervene in the networked circumstances which the imaginary influences today.

Taking into account the additional commentaries collected in this volume, our aim here is to clarify the conditions, to point out the underlying assumptions, and to question all these propositions. Musso's declared aim is a critique of the network and a constructive contribution to a broader understanding of technological innovation in the context of contemporary society. What does that critique really mean in his work, and does it match his advocacy of the imaginary revealed in his later writings?

Where is Musso coming from, and what does his discourse presuppose? What are the foundations of his reflection and, in particular, of the critique he proposes? Does the author escape from the circumstances on which his analyses are based by, for example, endorsing the discourse of technological innovation? Does he advocate a more enlightened symbolization of the network figure, and sustain the possibility of planning the technological imaginary, in order to meet the requirement for a critical project which frees us from those categories or, relying on the philosophy of the subject, does he in fact promote the empire of technology and revive its mythology?

Undoing the Fabric of the Network

Lucien Sfez states that communication "has become an important 'symbolic figure', that tends to unify the fractured social body, but also the scattered sciences, too specialized, light years apart one of other" (Sfez 1992: 41). Musso sanctions and enriches this perspective in observing that this communication, previously seen in linear or expressive terms, is now, in the context of what the first author calls "tautisme" (Sfez 1992: 99–114), defined in reticular terms. The author stands out among the critics of our technological condition in particular for his research into the historical roots and political trajectory of the network as a social fact established in the course of the nineteenth century by the heirs of Saint-Simon, at least in France.

For him, the imaginary is at work in the movement of technology, and its relationships to the real and the symbolic must be made known, if we are to acquire a critical understanding of our time.

> The network imaginary is this second-order dimension brought to light with the proper history of the word: on the one hand, it refers to something that captures and holds, like the fisherman's or the gladiator's net, the *retiarius*; and on the other, it refers to something that accompanies or allows the flows circulation. This is an ambivalent imaginary, associated

with full control and widespread circulation, uptake and relationship at the same time (Musso 2012a: 12).

The author argues that it is necessary to differentiate between the network as a specialized technical object and the network as the object of an ideological practice. The first would have philosophical and conceptual dignity, being linked to highly specialized scientific and technological research. The second would come from those people who, ignoring this aspect, exploit the figure with political and economic objectives, submitting it to a heteronomous dynamic.

Generally speaking, everyday experience confuses one with the other, as result of their common origin in a symbolic operation. This fact explains a history in which the network metaphor has taken on an epistemic function, but at the same time has been fetishized, converting itself in an ideology (*retiology*). The network has become an agency of technological innovation processes and, therefore, of institutional modernization. But at the same time this figure has come to be treated socially as a new relationship matrix and global vector for change, embodying a utopian fantasy from which we can only expect damage to science and disappointment for mankind. That is our problem, according to Musso.

For several years, people have witnessed the spread of a discourse according to which we live a technological revolution, based on communication informatics, one which will bring about a radical transformation of man, culture and society. Developments in information technologies drive a change in the structure of social life that, in turn, affects and feeds these technologies through ever more innovative applications. The technological changes that were a result of social processes in past times have now become the change in itself, in the eyes of contemporary man. An explanation for this might lie in the fact that these processes operate in a network and, from this, some analysts think we are entering into a new form of society, the network society, the name given to it by the Spanish sociologist Manuel Castells.

For him, let us recall, communication technologies help to overcome the social atomization deriving from earlier, more widespread conditions, without giving up the individualistic interaction protocols which have emerged in modern times. They are rearranging it into decentralized networks. The Internet relativizes the individualism of our age, making it possible to interact socially in a network. In this way it cannot be seen as a mere technological artefact for the exchange of information. The Internet has become a dominant form of social organization, to the extent that it reconfigures the trends to individualism released by modernity in a socio-technical network of global scale and impact.

The author adopts the liberal premise that "individuals," supported by the technical possibilities provided by computer communication, "build their networks, online and off-line, on the basis of their interests, values, affinities, and projects" (Castells 2001: 131). He advocates that individuals rebuild their social interaction patterns with the support of computer-mediated communications to create a new form of society: the network society (ibid: 133).

Castells is well aware of the dialectical principle according to which individuals and therefore their actions are structured more or less collectively. However, the fact that he refers the case to the historical and social context in which the Internet has agency does not prevent him from concluding that the other-directed individualism seen by Riesman in the late 1940s has been brought to a head by the Internet. He just resets this in a way more attuned to the selfishness dominating at the turn to the present century. "The Internet provides an appropriate material support for the diffusion of networked individualism as the dominant form of sociability" (Castells 2001: 131).

For us, it is clear that this approach, transcending technological determinism, realizes only one aspect of the phenomenon. Individualism is undoubtedly at the root of the development of interaction patterns such as those that form networks. This historical phenomenon of individualism, it is worth recalling, is the origin of the modern notion of society. At same time, however, we have to observe that society and networks, not to mention the individual and technology, are historical creations we need to relativize.

With this we say that there are two alternative ways of developing the argument about networks without succumbing to technological determinism. The first consists of deducing the network from the changes in the patterns of relationships between individuals in a given time. The second consists of going further, by explaining how this idea has been invoked by social beings to structure their individual relationships. Reference to the network society does not only account for a supposedly higher form of society, it also imports an ideological prejudice. It covers up the fact that, for many, this noun has now occupied the position which social order, social structure, and social system had occupied up to a few years ago as defining aspects of society.

The greatest merit of Musso's work is that he clarifies key steps in the process leading up to this situation. Contemporary thought tends to speak of networks as a social fact which was imposed with technological development. Thomas Hughes (1993) tried to avoid this approach by showing us how technical networks have developed under political and economic influences. Bruno Latour (2005) expanded this view by creating a network ontology disguised as a new discourse on method. Musso relativizes the metaphor historically to expose how it became possible and, more broadly, ended up becoming a new ideology in our time.

In France, his works assure us, we owe to the heirs and successors of Saint-Simon the creation of this social imaginary which emerged around the notion of network (Musso 1998, 2007a [2003]). Saint-Simonism is the starting point for the appearance of the network as a figure or metaphor for contemporary technological thinking. Beginning with it, "technological network means democracy, circulation, equality" (Musso 2007a [2003]: 127). Saint-Simonists were among the first to outline this world view, advocating the construction of transport and communication networks as a factor in economic progress and social welfare during the nineteenth century. "They saw in a complete understanding of technology the solution to a political and social crisis", and believed that in this way it would be "possible to reshape society thoroughly, not in Bensalem, but right here, in Europe", summarizes Jean-Yves Goffi (1996: 49).

Saint-Simon conceived our organism as a physiological network and, casting the metaphor over the social field, preached the political convenience of developing it technologically. "We must excite with the lure of specific benefits all companies whose purpose is the construction of canals, roads and bridges" (Saint-Simon 1859 [1822]: 443). His disciples and successors devoted themselves to exploring this latter aspect. According to them, if "the network provides the natural life of the organism", it is worth believing that artificial networks can fertilize and develop political and technically a geopolitical territory and its population (Musso 2007a [2003]: 122).

> The multiplication of railways on the continents and of steamships over the seas will be an industrial and political revolution at the same time. Through this and other modern inventions, like the telegraph, it will be easy to govern the major parts of the continents bordering the Mediterranean (Chevalier apud Musso 2007a [2003]: 125).

Saint-Simonists were thus the intellectuals who, in the mid-nineteenth century, sought to translate into a political and material project the moral and intellectual work of their mentor. Michel Chevalier stands out from the group, and was notable for paving the way for the ensuing reification of the notion of the network. This French philosopher and writer advocated that it could and should be operated objectively by the government and private enterprise, with a view to the full development of the wealth of society. For him, networks are the medium which will enable the circulation of wealth and encourage human relationships to develop.

> Improving communications is the same thing as working for real, positive, and practical liberty... it means achieving equality and democracy. The developing of means of transportation produces the reduction of distance not only from one point to another, but also from one social class to other (Chevalier apud Musso 2007a [2003]: 126).

There was a time when Church and State played the role of being the unifying reference for social life, by definition unable to cope with the reality of its infinite divisions, according to Musso's interpretation. Nonetheless, the advance of capital, supported by technological progress, has led to their gradual weakening. *Retiology* may be seen, in part, as the product of a collective operation by which successive generations of public intellectuals have tried to develop a symbolism capable of maintaining the social bond amid this growing circumstance. "Retiology triumphed as a contemporary ideology thanks to technical determinism, which was opposed to symbolic determinism [formerly associated with Church and State]" (Musso 2007a [2003]: 230)

The medieval Church tied social life symbolically to the divine and transcendent sphere. The modern Republic worked for their transition to the immanent sphere of human life. *Retiology* is a technology of the spirit, or a technological ideology that arises with their consecutive decline and, thus, the social disintegration of collective experience promoted by modernity (ibid.: 148). The Saint-Simonists laid one of its foundations, by suggesting that social progress "is ultimately based on the multiplicity of communication networks only" (ibid.: 127).

For them, the solution to mankind's problems is technocracy, and lies in the control and technical management of our circumstances. Social transformation needs to move from the political plan of action to the technological level of implementation of engineering works. Transport networks gradually mechanize it, making it take on the features of a natural process in an artificial organism. The emergence of railways and, after that, of the telegraph, telephone and electrical systems encouraged the development of this interpretation, which assumes the character of a true *retiology* under the impact of computer communication in our lives today.

In this way, however, "triumphant retiology represents the climax of the catastrophic commodified degradation of the symbolic and conceptual value of the network" (ibid.: 234). Converted into the means of redemption of man and society, the term goes on to become a technological concretization of a utopia that, whether to be challenged or, more commonly, to be promoted, engages all of us and spreads the belief that we are all working in the same way for the construction of a new world.

Lucien Sfez (2002a: 268–277) argued that, although it promotes technological development, the network has little symbolizing power. Its fragmentary and abstract character renders it incapable of unifying the imaginary, exposing its fragility as a way of maintaining the collective consciousness. For Musso, on the contrary, the *retiology* to which it gave rise performs a social function, even while degrading the symbolic and conceptual value of the concept in the sphere of science and technology (Musso 2007a [2003]: 234). We are dealing here with a phenomenon that acts as agency and, thus, may come to replace political authorities as a unifying force of the social imaginary at a time when traditional powers are disintegrating.

Assessing Musso's Reasoning

Turning to our analysis, we must observe that the author moves in the sphere of dialectical thought. The question which interests us is the extent to which this dialectic is a critical one. We will argue that it is not critical at all. Criticism developed in Musso's early writings did not rule out the possibility of overcoming the problem diagnosed in relation to the figure of the network. His later work offers us an ideological way to solve the problem by appealing to the notion of innovation.

Explanation of this idea requires preliminary clarification. In considering Musso's work, we will distinguish between two possibly overlapping stages. The first stage corresponds to the studies on the network as a symbolic form and ideology. The second stage corresponds to the studies on the role of the imaginary in the development of technological innovation. The mediating reference in both cases, despite the different contexts, is Saint-Simonism, a kind of metaphysics of modernity, as Martin Heidegger might say.

We mean by this that in Musso's studies Saint-Simonist thinking is more than a mere object of analysis. Criticized as ideology, Saint-Simonism also operates as a principle of the intelligibility of his research. A cursory reading of his work may give the impression that the author condemns network discourse as mere ideology, but this is not in fact the case. Musso not only does not reject the relevance or pertinence of the network as a symbolic representation of the truth, he also fails to distance himself from technological thought, from the belief in technology which has become established with modern times. His work is limited to challenging its ideological exploitation, the mystifications it promotes publicly and, accordingly, criticizing the negative impact it has on technological development.

For us, Saint-Simonism can be viewed as an expression of the metaphysical belief in those developments, a variant of the technological thinking instituted by modernity. We may speculate that this way of thinking also pertains in Musso's reasoning, even though he criticizes it. This remission reveals two aspects. The first is its philosophical debt to positivism in general. Musso sees reason as an autonomous principle which can be accessed to institute a discourse capable of clarifying the real and eventually bring about its reorganization. The second is a subversion of old-style positivism. The older positivist claim, according to which the real can be accessed immediately, setting aside irrational factors, is wrong or untenable. Reality is always mediated by the symbolic and the imaginary instances of existence. Thus the concrete unity of these aspects should not be ignored during social research.

Positivism from Comte to the Vienna Circle aimed to dispel myths, replacing unrealistic and irrational fictions and images by empirical knowledge. Propositions without factual support or experimental basis should be cut off from the field of true science. Human thought can and should gradually be released from fantasy and superstition through rational and experimental research. If science has a critical accent, it is based on this aspect.

Musso contests this doctrine by arguing that, at least for technology, it does not apply. Technology is a sphere inside which phenomena have a double identity. In this area, the real is presented at same time as the functional and fictional object. That is to say, we must maintain our awareness of the ideological discourse of networks, and take care not to be trapped in it. To think properly about them we must avoid falling into *retiology*. However, this is not to deny the appropriateness and usefulness of the network idea in research and scientific and technological development. Although the terms he uses are different, the author agrees with other scholars who state that "The word 'network', before or beyond the ideology it conveys, is a scientific concept [too]" (Parrochia 2001: 8).

As he says in this volume, "there are two dimensions to the network: one technical and the other technoimaginary" (Musso, Chap. 2). But at same time he adds that they interact through symbolic operations. The network idea may be applied to landscapes and territories as well as to the mathematical and informational fields of science. The first aspect, the technological in itself, is structured and develops rationally according to logical and experimental principles. The second refers to the representations that allow the first to appear and make sense for their social subjects, whether or not they are experts. Technology is made of instrumentality and representation, the author sums up.

The effect of this is that although technology is increasingly influential, it does not have the autonomy that its own social subjects often claim, or the power that the masses tend to give it. Technology-mediated transactions between science and everyday life impose restrictions on it, determining its limits. Political power interferes with its development not only in official institutions but, through legislation and management of investment funds, in private enterprises too. For the most part, technology usually benefits from this, inasmuch as the dominant circles derive advantages from its development. The problem is that, because of this, it needs to be and in fact is mediated by and for social subjects, both in terms of symbolic articulation and in making sense for common people.

Musso concerns himself with the fact that this combination and this meaning tend to mystify. They overshadow the social awareness not only of the masses but of the technical and corporate leadership too. Unlike the older positivists, he knows, but in our view only up to a point, that reason cannot get rid of all irrational factors. The symbolic and imaginary elements are not ornaments or a traditional legacy which the sciences, politics, business, and society can eliminate, since they merge and necessarily interfere in their development. As a good positivist, however, he believes these factors, like any other, can be objectively and rationally handled. Politics not only interferes and places controls on them, but is capable of sustaining its rational management according to a will which is simultaneously more aware and voluntary.

> In other words, critical theory is a principle of analysis that acts on several records: it identifies ideologies, unmounts utopias, produces methods and concepts... but it is also an invitation to the manufacture of antimyths. It separates carefully metaphors and concepts, but it also knows they remain inseparable, like day and night (Musso 2006: 431).

Musso thus rejects technological determinism, and criticizes the symbolic operation that in promoting the figure of the network converts it into an ideology. But he does not cease to believe in technology, which he sees as a systemic and indelible force of our age. He wagers that we will be able to free it of irrational exploitation, to respect its limits, and finally lead it to a more enlightened political use. Proof of this lies in his support for business intervention in the field of the imagination, as we may see in his more recent work. Following a kind of late humanism, criticism in his work stems from a peripheral belief in man, in which admission of the imaginary element, naïvely denied by more narrow-minded positivist and technocratic thinking, is nonetheless subjected to the rationalist assumption that, albeit in a non-essential way, it may be politically managed with eudemonic goals and an enlightened attitude.

The analysis of what he has called the symbolic economy of the network does not deny its necessity as a concrete expression of the real world, and its relevance as an instituting fiction of everyday consciousness within and outside technological environments themselves. Reflective thinking confronts solely the politics that remains irrationally unaware of history and blind to the benefits that could result from a more educated point of view, and operates and maintains a deplorable state of affairs which could be overcome through renewed spiritual mobilization.

We are dealing with the problem that, "from one side, we are explaining the social whole by the network, and this is fetishism since it takes the part for the whole. But on the other side, we reduce the network to an architecture, which is fetishism again or at least technological determinism" (Musso 2012a: 15). Nonetheless, the network is a rich and useful concept for recording our circumstances, provided it is rationally controlled and managed with clear and strategic purposes in specific fields by responsible professionals.

Musso understands the symbolic and the imaginary as anthropological data from a phenomenological consciousness whose architecture is topped by the principle of sufficient reason in the way it was instituted at the dawn of the modern era (Heidegger 1991 [1957]). "Humanness symbolizes the same way it breathes", said Pierre Legendre (1992: 27). Musso welcomes and endorses this view, originating in Lacanian metaphysics, via a reception of the writings of Lucien Sfez. Along those lines, he states the principle that existence is structured in the registers of reality, of the imaginary, and of the symbolic (for a deep discussion of this problematic, see Lefort 1979: 295–345).

The network is an image of the real, the ontological status of which is both symbolic and imaginary. The symbolic character comes from the fact that it is the product of collective operations and discursive practices that define this record. The network is a prosthesis historically constructed by practical and discursive procedures, through which the real is revealed to consciousness and thought. The fictional aspect has to do with the fact that intentional consciousness is not enough to articulate it, and thus requires sensitive materials and plastic creativity to impose itself in everyday life.

Social subjects, argues the author, intervene politically, that is, in modern terms, under conditions that presuppose their representation as a collective, but in fact implement that agency in a contradictory and conflictive mode, due the diversity of groups and social classes. Despite its divisions, society tends to be represented and managed by each of its parts as if it were unified through the creation and subsequent exploitation of symbols charged with universal appeal. The decline of religious symbolism that supported a relatively more consistent social unit in the medieval world was accompanied by the emergence of a politics based on secular symbols, the result of which, after a time, is the creation of political symbolisms, as taught by Sfez (1988).

Advances in technology, and the increasing dependence of political power in relation to it, add a new feature to this situation insofar as they represent a growing colonization of these symbolisms by images attached to them. The research and technological developments that now impact on the field of power begin to be fictionalized by an imaginary activity. The network finds expression as a broader social metaphor in this context, inasmuch as it has shown itself to be capable of mediating the spheres of politics and technology.

The problem is that this fact also paved the way for its ideological exploitation and, therefore, for the legitimation of a new power system in apparently postpolitical terms.

6 From the Critique of the Network Symbolic Form to the Ideology of Innovation...

> On the one hand, upstream, there are engineers who invent products and techniques, 'sociologist engineers' in the words of Michel Callon, since they think about the social applications of their technology, about the virtual, fictitious, and imaginary uses and users of this technology.
>
> On the other hand, downstream, looking at the user, there is 'the creativity of ordinary people', as shown by Michel de Certeau. It is hidden in a tangle of quiet and subtly effective wiles, by which each user finds a way to walk through the forest of products imposed by industry. The user will choose, adapt, and subvert the possibilities offered by the technical object.
>
> In between this, there is a multitude of mediations on which it is important to act to ensure the social and cultural structure of the service. One can speak about this as an 'abundance' of imaginaries, but always in a framework that is shared by all at the same time (Musso 2005: no numbers or paragraphs).

For us, this passage reveals the way that the Lacanian metaphysical framework in which Musso develops his discourse supports and is sustained by another equally comprehensive and fundamental context, in which the subject is the primary point of reference and its being is determined in its essence by the figures of power and reason (Heidegger 1991 [1957]).

Musso notes, and we believe correctly, technology's need to be symbolized, although as far as we know he does not clarify how he conceives that concept. Modern technology itself does not create or at least does not keep us in social relationships, given its highly selective and esoteric character. We need it be symbolized, i.e. to be expressed in common language so that it is integrated into everyday life, whatever the social context. For this reason, as already mentioned, Musso engages in polemics with the ideological exploitation of network symbolism, and not with it *per se*. He sees it as an anthropological datum constructed throughout the course of our history.

Why he does this is not clearly explained – and this is significant in terms of the arguments adopted here. We have observed that his major works circulate epistemically within the limits of dialectical thought and of the critique of ideology as originated in the Frankfurt School (see Geuss 1988). The advantage of this is that it does not reduce the concept of ideology to a body of ideas, as these ideas can be experienced by us. The term applies here to the logic that controls this corpus, in spite of the justifications that the social discourse responsible for its invocation may provide (see Hoy and McCarthy 1994). The difference with the Marxist formula lies in the fact that, while in the latter the term is dialectically connected to the concept of the idea, in the present case we find its epistemic counterpoint in the concept of the symbolic form, extracted from the works of Legendre (Legendre 1983, 1992, 2007) and Sfez (1988, 1992).

In overall terms, adoption of this method means that the author avoids a radical confrontation with the topic under analysis. Analytical reflection on the matter preserves its moment of truth as much as the possibility of transcending those aspects he has diagnosed negatively. In Musso's work, rejection of technological determinism does not require ignorance of the fact that technology has become a determinant of social change after the Industrial Revolution. Nevertheless, the lack of a stronger reflection on this point reveals the dependence of his analysis on technological thinking, as we may see in a full reading of his writings on innovation, but also when investigating more closely his earlier work on the network idea.

For him, the damage attributed to technology is not technological. It would be based on an ideological procedure which, operating according to the logic of power, causes technology to break its own limits, and so converts it into mystification. The problem with the network does not lie in the idea itself, but in the exploitation that has transformed it into a kind of general matrix, if not the essence of our society. Important voices made it a key to the solution of major political and human challenges of our time, inflating a "metaphorical use of the network that contributes to the final condemnation of the notion in itself, as if just its abuse were not enough to determine its loss of meaning" (Musso 2007a [2003]: 199).

As in every ideology, the network idea causes embarrassment when it serves to project the part over the whole and, thus, confers to technology a capacity for social change that it does not in fact have. The formation of technological networks is accompanied by myths and images which enable their socialization. But with this they simplify the collective utopias, and reduce politics to misleading fictions, insofar as they are presented as a form of direct, instantaneous, participatory, and universal democracy – as an anti-hierarchical, anti-bureaucratic, and anti-pyramidal social fact.

The result of the deficit of meaning that engenders a society increasingly dependent on its technological and rational institutions is a reification of its symbolism, the author notes, although he himself sometimes falls into the discourse he is questioning, as we may see, for instance, when he states: "today there is such a speed of technological innovation, such a strength, such an invasion of technology in all social practices, that there is no need for manipulation" (Musso 2012a: 15).

By virtue of this, *retiology*, the contemporary ideology of the network, should not be seen as a form of manipulation, but as a social fact whose sources come from all sides and are collective in nature, he notes in his interview with *Culture Mobile*. "There is certainly a naturalization of technology, but this fact, this power of the network as a metaphorical description of our society depends primarily on how our society is increasingly ruled by technology" (ibid.: 15).

The conversion of the network into myth, and its social presence as mystification, is liable on conviction, but this does not mean the notion lacks an objective function, as it allows us to symbolize a process that really is transforming our society. We cannot live socially without a symbolic mediation of reality. Saint-Simon was perhaps "the first network and network society philosopher", but not only as an ideologue: "The industrial society whose shape he delineates before its time, [really] is a fluid and networked society, a society marked by a widespread circulation [of information], in which networks are so multiplied that they have ended up encircling the planet" (ibid.: 19).

Musso makes this idea explicit in a more recent text, affirming that 'deciphering the ideology of the network does not mean criticizing the use of network technologies. Rather, it is a question of shedding light on them, without confusing technology with the social-political realm' (Musso 2012b: no numbers or paragraphs). Although he is clear in his critical attitude to *retiology*, inasmuch as it transforms the network into an ideology, he also recognizes the relevance of the concept as a symbolic element that interferes with technological innovation. Its ideological meaning lies only in "the widespread and abusive employment of the notion [of network]" (Musso 2007a [2003]: 199).

As we are arguing, the problem with this reasoning is that it causes his work to collapse into a mitigated form of technological thinking.

Marc Guillaume observes that 'the multiplication of channels and information networks has the merit of reducing the importance of the central public scenes, which exercised a conservative and reductive power in the past'. Moreover "we must recognize that networks have allowed the appearance of new internal mediations of civil society, a kind of 'electronic word of mouth', that partially offsets the destructuring effects provoked in it by the traditional mass media" (Guillaume 1999: 145).

Musso follows this line of analysis, making clear that the objection to ideological abuse of the fiction represented by the network as much as to its mystifying effects on the broader social context does not prevent us from asserting its functionality as a symbolic agency of the technological realm. The dialectical treatment of the subject is revealed in the reservation that, since it is a function of moral or political points of view, there are good and bad uses of the concept, even outside the environments in which it is legitimate. Social networks may provide support for movements opposing a dictatorship, but may also enable espionage by defenders of that dictatorship; in sum, they may promote terror as much as they may promote humanitarianism (Musso 2012b).

We may conclude that for Musso humankind is at least partly free from technology and yet may not totally dominate it. Some people are able to escape from technology and its ideological emanations. His critique follows the lines of a form of dialectical thought nurtured by a philosophy of the subject tailored in positivistic patterns.

As Sfez (2002a: 74–75) notes, political decisions incorporate technological networks to make them serve as agents of social change, but also to hamper or even to block such change. From here we may conclude that, if we understand the contradictions contained in the historical expansion of the network imaginary, we may be able to study and perhaps open up the possibility of political intervention to limit its ideological dimension and to explore its useful or productive aspects in a rational manner.

We suppose that this is the case for Pierre Musso, as we will try to demonstrate below.

Weaving a Paradox

Musso does not reduce technical progress to ideology, inasmuch as he recognizes its historical reality, nor does he deny the role imagination plays in technology. As a late and enlightened positivist, he admits that the imaginary has always been or, at least, has become a necessary condition for technological development. Technology itself does not symbolize social reality, and at same time experts and technicians are unable to develop its imaginary. This task is taken on by thinkers, writers, journalists, marketers, etc. Thus, the tasks we may take on as critics would consist of not only exposing their mystifications but also indicating, using a constructive approach, how their function may be rescued socially. The imagery can be worked out in favor of technological innovation, rather than promoting a mythology that, by exaggerating its meaning and scope in a good and a bad sense, promotes irrationality among the general public.

The network is not a fiction which can be set aside at will: it is a symbolic expedient by which technical activity is expressed in the imagination, since without its operations technology has no way of being managed among common people and even among technical experts as social subjects. The network has become a fictional image for personnel in the information milieu and for the scientific community, which is characteristically inarticulate as far as discursive skills are concerned, it has become what might be called a spirit technology (Sfez 1992: 377–392). Still, it tends in this way to reinforce its condition as a colonizing force on social consciousness, to the extent that technology is increasingly influential in everyday life.

What does this imply from a practical point of view? For Musso, there is no reason to succumb to despair and to consign the subject to its historical fate. The philosophical and political positioning that we need today is not that destined to destroy or to forget the notion of network. We need an attitude that reworks it rationally through the development of an affirmative action. Due to our objective historical context, this looks promising for the business environment only, particularly in the field of media companies (see Musso 2010).

In France, Loïck Roche and Thierry Grange (1999) showed signs of this understanding by starting to advocate the need to bring the imaginary into the heart of technology research and development business programs at the end of 1990s. For them, innovation processes have entered a stage in which the mere investment in computerization is no longer sufficient. We are now in an information society that requires a more holistic and complex vision of the subject. Human sciences have revealed not only the social resistance to technological innovation (Roche and Grange 1999: 69), but also the dependence that this fact has, as innovation too, on the imaginary element (ibid.: 76–95). So, they conclude, it is now urgent to understand that technology and imagination, rather than opposing each other, can and should converge. Both are human creations and, because of this, it is desirable that we deal with them as allies, if they are to grow well, as advocated previously by André Malraux (ibid. 169–170).

Musso takes up this point of view by arguing that, to design innovation in levels more attuned to the economic and business scenario emerging with the development of new technologies, we need to study the imaginary and to invest in its practical applications. Technological progress has impacted on the imagination to the point that it is now a sphere that is "increasingly complex and unfolds in the plural". The fact should not make us forget, however, that in this way technology has been influenced by the imagination at its very heart.

> Imaginaries inform and shape the technical objects in the innovation process. Technical objects are a social and cultural construction, whose imaginary one can read as a structure formed by many geological layers, and so turn into raw material to develop and analyze, or to modelize [in theory and practice] (Musso et al. 2014: 5).

Sfez (1988) has referred to symbolic politics as a remedy for the developing crisis of political representation inasmuch as modern times undergo and, in a subsequent step, succumb to technological influence. Symbolic politics, for him, means a set of operations and images that aim to maintain social bonds in a divided society that has lost its transcendental foundation. For Musso disclosure of their limitations, if not their failure at the current time, suggests we should use the concept of symbolic economy instead. With this, we may gain a space in which it will be possible to intervene positively and practically (Musso 2002).

Analyzing Musso's work, we have noted that genealogical mapping of the network as symbolic representation of the real seems to be, above all, a fight against its conversion into ideology. As we have said, the author did not say it should be abandoned, as the older positivists would have done; criticizing the mythical aspect and the ideological dispersion of the representation, he accepts the possibility of its enlightened use. This means that the standpoint he adopted to address the issue of the network is valid for the imaginary too.

> We must avoid two excesses in addressing this concept: on the one hand, the rationalist rejection of the imaginary in the face of the 'real', and on the other, the fascination that takes away all meaning from the concept. Following this path one could say that the imagination is at once a transformation of the real into a representation and a realization of representations (Musso 2007b: 13).

In the new stage of his research what seems to change is the axiological significance, not the epistemological scheme itself. Argumentation passes from the historical and political denunciation of *retiology* to the theoretical development of non-rational foundations for entrepreneurial action. Innovation processes rely on technological research but also on the myths, images and fictions that intervene in its development, invading and permeating the conception of products and services in industry. Technical staffs still tend to neglect this factor, but it is time their leaderships assumed the existence of the imaginary, considering it is "a raw material in the process of innovation" (Musso 2007b: 39).

In the past, there were those who, raising the flag of industrial design, defended this idea with a predominantly technocratic attitude. In Germany and America, the pioneers of the technocracy of the senses and the manipulation of the imaginary have proposed an aesthetic revolution that would incorporate the mass mind into the patterns and rhythms of corporate capitalism (Buddensieg and Rogge 1984; Ewen 1988; Noble 1977). After World War II, postmodernism overcame this phase, to indulge in a strong eclecticism of opportunistic motivation, in which the aesthetic form ended up subjugated to the immediate reactions and stimuli induced by the social tendencies inherent in a reconstructed capitalism (Subirats 1986; Sennett 2006).

Musso revisits this issue in a more academic key, in an effort to conjugate the old modernist expectations for a rationalization of culture with the ever more arbitrary, unexpected and liquid desires of the social subjects of our time. Like Bernard Stiegler, he believes that we can make difference by intervening in innovation processes. "Our dream is to venture boldly into the process of industrialization of the technological imaginary, which never stops to change, in order to innovate and to create in and on the innovation" (Musso 2010: no numbers or paragraphs).

Nonetheless, their points of view are not the same. "Inside Ars Industrialis, we advocate the reconstitution of a real public policy [...] developing an investment in the mind that cannot be done solely by capitalism", says Stiegler (2008: 93). Musso, by contrast, seems to favor the innovation policies of the business world. For him, contemporary corporations need to understand consumer expectations and their interactions, but also and at same time to provide an effective and creative response to their challenges (Musso 2007b: 9–62).

Corporations today must not ignore the power and influence of the imagination. Entrepreneurial activities have become a link between industrial routines and knowledge applied to the production of wealth. Research on the imaginary opens up the possibility for their managers to approach in a more rational way the imaginary that pervades them as much as the life of their customers. To design products and services that consumers will adopt, it is necessary to make room for the imagination. "If we want to sell the dream, it is necessary to know how to manage the imaginary of our customers", as Bernard Charles states in a postface to the book we are referring to here (Musso et al. 2007: 283).

From the work of Sfez, the author takes the teaching according to which:

> There is a community of beliefs and practices, which fiction must subscribe, if it is to be credible, if it wants to be heard. The same is valid for technological innovations: they must correspond to a feasible option for people. We must be able to understand them in the same way that fiction must be credible. Technology has to go through a kind of consensual approbation, by the test of a common use, to function and achieve its goals with success (Sfez 2002b: § 22).

All this leads us to conclude that Musso's work is in part criticism, in part strategic and instrumental thinking. The historical research into networks has revealed their transformation into ideology, with prejudice to technology and our political consciousness. The applied study of the imaginary can help corporations to manage the circumstances influenced by them in a more useful and efficient way. Companies are now being called on to design goods well-fitted to the needs of a new, post-Fordist consumer if they want to thrive. At same time, however, competition has become global, and social networks tend to cause dispersion and arbitrariness in the attitudes and preferences of all kinds of customer.

Scientific shaping of the imaginary relevant to business, and its infusion into technological developments, can help corporations to promote a convenient strategic adjustment for both, projecting innovation to a more advanced and enlightened level. Their managers cannot proceed in autocratic and anarchic ways in relation to this variable any longer, since the pace of business communications is increasingly determined by the idiosyncrasies of consumers too. The cultural planning of their actions and the shaping of the imaginary of their clients to their innovation policy have become a survival factor for their managers, argues the author (Musso et al. 2007: 28; cf. Musso et al. 2014).

Musso draws on Sfez (2002a) to highlight the contrast between those who work with technology and those who talk about it, observing the skills that the former should provide to avoid the damage to technology from political exploitation of the word and, above all, of images. This perspective reveals the way in which the author sustains that it is possible to articulate network discourse with less ideology, following a line that was already present in his early work.

We have seen how he moved away from old-style positivism by recognizing the inevitability and the influence of the imaginary on human life. Providing it with new means of exploitation, industrial society not only reinforced this factor but developed an imagination apparently still more radical than the previous one. Now, the challenge facing their (rational) subjects is, in managerial terms, to find out how to regulate it to their advantage, without ignoring the public; it is to know how to turn to their own advantage the mythologies, dreams and images which impact negatively on their activities, sometimes with catastrophic effects.

According to the author, the answer to this lies in an effort to attune technological innovation policy to knowledge of the social imaginary. From now on it is necessary to develop a business strategy which preserves technical rationality, sustains economic competitiveness, and responds reasonably to the wishes of a complex cluster of customers.

> [In short] It is guided by two key ideas: 1) the imaginary is the 'raw material' – susceptible to formalization – of an innovation that has become intensive; 2) there is a field of research to be explored at the junction formed by the imaginary, creation and innovation, considering that the aspect of the perceptions, emotions, and representations is not yet well studied (Musso et al. 2014: 5).

Thus, he explores in his work on technological innovation a similar ambivalence to that encountered in his studies on the network. Musso believes that "there is a convergence of the two processes: from the 'artistic side', the integration of technologies (including IT) within the creative process; and from the 'industrial side', the claims to enrich the innovation process made by the artists" (apud Caraës 2008: 110). The difference between these two stages of the research lies in the fact that this scheme now has a political and doctrinal tone rather than a historiographic and critical one.

> Work on 'the imaginary', positioned upstream of the innovation process, can integrate and even anticipate their uses in a 'light' way, recurring to simulations and montages, even before the development of a prototype. At this stage of the process, the concepts are still malleable and rectifiable: later, economic or industrial constraints are such that the correction becomes difficult. When we work on stories, narratives, and images, we can intervene in the process in a light way, before the restriction of possibilities. We also analyze the insights and perceptions interchanged in the global networks of collaborative work using

images and worlds of forms, notwithstanding the diversity of languages and cultural references (Musso 2010: no numbers or paragraphs).

Recognition of the homogeneity of procedure is a sign of the coherence of his approach to technology in our time. The point to note is that, in this second phase, the critical insight is missing. Accepted as a given with which corporations work, the imaginary is simply reified. The critique of *retiology*, as we have seen, did not exclude recognition of the functional network as a regulatory concept in techno-scientific research and as a symbolic operator of our time. To praise the role of the imaginary in technological innovation seems to grant it legitimacy:

> Today, the injunction to continuously innovate, and the transformation of technology into a veritable totem, situated in the middle of everything, forces us to 'naturalize' the network to the point that it becomes the dominant metaphor of our relationship with the world (Musso 2012a: 17).

Musso devotes significant effort to researching a new industrial imagery, but in so doing omits the critical reflection necessary to maintain intellectual consistency with his earlier work. He does not appear to realize that there may be an ideological rationalization of a process here that not only transcends this register, but reflects the will to power that increasingly drives the capitalist market economy. The focus of his investigation is purely strategic and instrumental, even if it recognizes and deals with an irrational element in our existence.

> Anthropology, sociology, psychoanalysis, semiotics, all the human and social sciences, showed the ability to detect architectures in these dialogues about innovation. The business of science in itself is to modelize. But if it is to modelize the imaginary, we have to apply logic to that which lacks it, concepts of a non-conceptual nature, formalizations, including mathematics, over 'what escapes us' (Musso 2010: no numbers or paragraphs).

When the capitalist will to power intervenes in the argument, it is presented as support for managerial research and thus gives it epistemological legitimacy. "The contemporary challenge is to explore the imaginaries created with the new artificial worlds we build daily in networks in which we act and in which we meet with others" (Musso 2010: no numbers or paragraphs).

Today corporations care about innovation because this is a strategic question of survival, necessary to their future in the midst of global competition for control over mass markets and the soul of consumers. Governments invest in policies supporting this because of its supposed benefits to the nation, but also because economic growth has become the way by which power is exercised and maintained. Finally scientists and engineers have an interest in it because we all expect its contribution to collective development, and they get more power for them with their efforts to develop technological innovation.

Since Musso passes over all of this, we have to ask whether, having criticized network ideology, he himself has succumbed to the ideology of innovation. We have to enquire if his academic work and historical reflection on the network, rather than being an authentic critical engagement with its problematical effects, is nothing more than a prelude to a political voluntarism obfuscated by that which he has called *retiology*.

From One form of Saint-Simonism to Another

Musso teaches us that already in Saint-Simon, "politics does not play a very important role. He reduces political power to managerial competence". If we consider his pioneering research into Saint-Simonist thinking, this fact becomes even more evident. "Politics limits itself to the diversion and crystallization of social flows: it must be transformed, or removed" (Musso 1998: 162). Later technocratic movements reinforced this cancellation of politics, since their supporters contributed decisively to spreading the ideology according to which there are scientific concepts and objective measures for solving our problems. As neoliberals and many retiologists argue today, good governments are those that manage public affairs as if they no longer exist. "If the goal is 'the development of the information society', it is necessary to 'reduce and neutralize the state through communication,' they say" (Musso 1998: 299).

Around 2009, Musso still had reason to criticize the triumph of "imaginary industries", pictured by him as symbolic and mythological expressions of the transfer of power from the political to the economic realm promoted by neoliberalism. He still thought that, underlying this transfer of state hegemony to the private sector, there was a negative ideological process by which the political tends to become invisible in our society (Musso 2009a: 242). Envisaged as an effect of a symbolic operation allowing the transfer of State power to private corporations, the 'communication society' would be a discursive prosthesis destined to camouflage the political fact, and thus also the original divisions of society. The author's concerns were with the processes that help to undermine the knowledge of our own reality (ibid.: 243).

In another article from the same period, he even questioned the "rationalization of the imaginaries" promoted by the new managerial tendencies, denouncing the "comm-management" ("*management communicateur*") as a "soft barbarism" that "acts for the necessary adaptation to the fact of globalization and of technological innovation" (2009b: 130). It is surprising how the author, without abandoning the critique of *retiology*, surrenders to an uncritical discourse, and promotes an ideological legitimation of innovation in his writings of recent years. The following passage, published in 2005, is an early proof of that, a speech celebrating popular participation in the business world which reproduces or seems to be extracted from the general public section of an annual CEO Address.

> Innovation has become a collective phenomenon, complex, large, iterative, involving designers, all of the company, its partners and its competitors, the media, and users who are no longer mere 'receptacles', but have become co-players (Musso 2005: no numbers or paragraphs).

We believe Musso shows how Saint-Simonism can be criticized without distancing oneself from it, which means keeping one's thoughts, as well as social action, focused on the will to power in modern metaphysics. "Sfez says that 'criticism is useful' but insufficient, considering that 'human beings need founding myths'. In sum, the only fight that works is one myth against another" (Musso 2006:

430). Nietzsche might well subscribe to this statement, taken up in an essentially administrative fashion by Musso. From a critical reflective perspective, is there another way of interpreting his purpose of forming classes engaged in processes of interdisciplinary innovation, skilled in creating industrial forms which can renew their imaginary in a way both reasonable and profitable? (see Musso et al. 2014).

We do not see any problem in an intellectual effort which takes a sober view of the imaginary, in any project approved for managerial use which aims to combat its disproportionate influence on the field of technology. Consistent and promising innovation is perhaps produced when the imaginary connects to (real and rational) knowledge, rather than being weakened irrationally by collective social fantasies. Innovation probably works better when it is not carried away by false aspirations, a source of frustration within and outside corporations, and when it provides balanced and convergent linkages between imagination and technology. Anyone who thinks in this way can be seen as an enlightened and thoughtful positivist, a rational person who does not rule out the imaginary in life, but instead puts it toward the synergy between companies and consumers, if he or she works for a corporation.

In the final analysis, who knows if this line of conduct is positive, negative, or indifferent towards technological development? Moreover, we may ask whether the claim to innovate by allying rationality with the imaginary is something reasonable and feasible, or is just an ideology that, in appealing to socio-technical networks, could be included in *retiology*. Up to what point can we deal with the imaginary as a sensible set of representations and, considering its sources and mode of being, as something that can be successfully shaped instrumentally by the knowledge of a competent subject? Does this idea not contain a neo-Saint-Simonism, based on the alliance between the academy and big businesses, whose rationalist claims are, in the light of the way global life has evolved, largely fanciful?

Regardless of the answers, we believe that the underlying endeavor is doomed to failure. Nobody doubts that corporations are getting results through technological innovation. But in overall terms these results cannot be explained solely by their postmodern rationale. The entrepreneurial intention of shaping the imaginary is a power fantasy. The further it is projected, the more this fantasy, i.e. imagination in itself, vitiates the will to power underlying administrative research into technological innovation. In fact, it is the advance of this will to power over the whole of existence, and not the supposed knowledge we might have of the imaginary, which explains the maniac search for innovations and their eventual adoption, in an age in which virtually all of us are more or less under the spell of the will to power.

Musso would have been more consistent with his earlier work if he had submitted his later work on the imaginary and innovation to the same line of reflexion and analysis he had applied to the network phenomenon, and if he had questioned these categories as a form of symbolic frameworking of technological praxis and an object of ideological discourse within corporations in contemporary capitalist society.

At least, he could have observed that technological innovation is a function of economic competition, as noted by Schumpeter and restated recently by Stiegler (2010: 81–103). That it serves as agency for a series of actions and constraints, but is not always successful because now, as before, there are many people who ignore or resist them. Apparently Musso interrupts his analysis in the first phase of the innovation process, the time of prophecy and fantasy, as established by Scardigli (1992). Refusing to accept as a practical principle the fact that there is a second phase, of failure or deviation, as Scardigli says, he wants to overcome the circumstances that lead to the problem. This leads him to forget that, despite men having entered the technological world, they do not control it as they want or think they do.

The only lesson Musso draws from Schumpeter's work (McCraw 2012) is the strategic relevance of the concept. He omits the morally negative effects condemned by the British thinker. He does note the anxieties and institutional impasses that corporations and their technical staffs experience in the midst of an economy marked by uncontrolled expansion from the standpoint of the ideological tranquilizer that the scientific administration of the imaginary represents. But he fails to draw a critical conclusion from the fact that innovation, whether technological or other, was not seen as a blessing until the end of nineteenth century. Besides, he seems to ignore the circumstances in which 'the contention that societies must have official policies dedicated to promoting scientific research and technological innovation was formulated after the end of Second World War' (Leiss 1990: 24).

Against this, we think that it is essential to see that innovation has not been widely used as a political keyword within corporate and governmental milieus before the modern day, as has been demonstrated by Benoît Godin (2008, 2014). Formerly innovation was seen as a negative aspect of social life. Francis Bacon, to cite a relevant voice, saw in experimental science a strategic tool for humanity's future. With its help, he believed, the false hopes and old illusions from the past could be supplanted, with man entering on a new epoch. Nonetheless, in his own time people did not yet see innovation through a positive lens. "Except the necessity be urgent, or the utility evident", innovation was a kind of thing that we had to ponder with gravity before putting it into practice.

> And well to beware that it be reformation that draweth on the change, and not the desire of change that pretendeth the reformation. And lastly, that the novelty, though it be not rejected, yet be held for a suspect; and, as the Scriptures saith, that we make a stand upon the ancient way, and then look about us, and discover what is the straight and right way, and so to walk in it (Bacon 1906 [1597]: 74–75).

Understood as the production of new goods, ideas or services to be brought to market with the help of technological research and development to bring about changes in society, innovation was yet to be implemented as entrepreneurial policy at the beginning of the twentieth century. In the early 1960s, the category was still not widely accepted. For economist Machlup [1962], "we shall do better without the word innovation", as Godin has documented (2008: 35). Only in recent years has the belief that technological innovation can change social relations and create a new world become a hallmark of contemporary government, a focal theme of public discourse, and an element in the spiritual climate of our time (see Morse and Warner 1966).

Musso passes over all of this, but in so doing tends to confer academic legitimacy to the postmodernist managerial postulate according to which innovation now demands an exploitation of the imaginary, a shaping of its plastic movements, if it is to be successful and its marketing potentialities are to be maximized. He develops the thesis that its technological substance may be a mechanical design, and has a direct, connected effect on social relations. Instead of submitting this fact to historical and reflective critique, he argues that the growing ability of social groups to articulate their interests and to express their tastes creates a new variable between businesses and their customers.

With this, he does not see that, whether or not it appeals to the shaping of the imaginary, technological innovation has become an ideology in the same way that networks have, and that there is no way of overcoming this fact by summoning up the intervention of an enlightened technocracy, at least in the midst of the anarchic economic development that has become planetary in scope. We may conclude that author's network critique ends up as an acritical elegy and a propagandistic endorsement of technological innovation as a progressive force and element of social redemption. Technological innovation appears in place of social change in the midst of an era in which political utopias gave way to consumer satisfaction in a way similar to how the network does this to organized, teleological, and concrete political and social movements, as shown by the Musso himself before his turning into a prophet of technological innovation and the shaping of imaginaries.

Confirming his indebtedness to the philosophy of the subject and its customary anthropology, Musso proposes a proactive understanding of technological innovation that does not ask whence it came, what its limits are, and how it matters – if it really matters beyond the motivational speech sustained in the business world, when it mingles with other discourses and images in the processes of practical and ideological subjection occurring in the contemporary world. Although he reveals and warns us of the mystification involving a category we have used blindly to understand and articulate our circumstances in his important writings on networks, his work on the connections between innovation and social imaginaries leads us to think seriously about the limits of a dialectical critique of ideology in an era dominated by the will to power brokered by market mechanisms.

References

Bacon, F. 1906 [1597]. *Essays*. London: Dent.
Buddensieg, T., and H. Rogge. 1984. *Industriekultur*. Cambridge: Cambridge University Press.
Caraës, M. 2008. *La création artistique en zone de marnage: Actes des Ateliers sur la contradiction*. Saint-Etienne: École Nationale Supérieure des Mines.
Castells, M. 2001. *The internet galaxy*. Oxford: Oxford University Press.
Castells, M. 2013. *Redes de indignação e solidariedade*. Rio de Janeiro: Zahar.
Ewen, S. 1988. *All consuming images*. New York: Basic Books.
Galloway, A., and E. Thacker. 2007. *The exploitation: A theory of network*. Minneapolis: University of Minnesota Press.
Geuss, R. 1988. *Teoria crítica: Habermas e a Escola de Frankfurt*. Campinas: Papirus.
Godin, B. 2008. *Innovation: The history of a category. Project on the Intellectual History of Innovation*. Working Paper 1, Montreal.

———. 2014. *Innovation and creativity, a slogan, nothing but a slogan. Project on the Intellectual History of Innovation*. Working Paper 17, Montreal.
Goffi, J.-Y. 1996. *La philosophie de la technique*, 2nd ed. Paris: PUF.
Guillaume, M. 1999. *L'empire des réseaux*. Paris: Descartes & Cie.
Heidegger, M. 1991. *The principle of reason*. Bloomington: Indiana University Press.
Hoy, D., and Th. McCarthy. 1994. *Critical theory*. Cambridge: Blackwell.
Hughes, T. 1993. *Networks of power*. Baltimore: John Hopkins University Press.
Latour, B. 2005. *Reassembling the social*. Oxford: Oxford University Press.
Lefort, C. 1979. *As formas da história*. São Paulo: Brasiliense.
Legendre, P. 1983. *L'empire de la verité*. Paris: Fayard.
———. 1992. *Leçons VI*. Paris: Fayard.
———. 2007. *Dominium Mundi*. Paris: Mille et Une Nuits.
Leiss, W. 1990. *Under technology's thumb*. Montreal: McGill-Queen's University Press.
McCraw, T. 2012. *Joseph Schumpeter e a destruição criativa*. Rio de Janeiro: Record.
Mercklé, P. 2011. *Sociologie des réseaux sociaux*. Paris: La Découverte.
Morse, D., and A. Warner (eds). 1966. *Technological innovation and society*. New York: Columbia University.
Musso, P. 1998. *Télecomunications et philosophie des réseaux*. Paris: PUF.
———. 2002. L'economie symbolique de la societé de l'information. *Revue Européene des Sciences Sociales* 40(123): 91–113.
———. 2005. *L'imaginaire au service de 'l'innovention'*. Aix: Fondation Internet Nouvelle Génération.
———. 2006. Réflexions sur la théorie critique de Lucien Sfez. In *Politique, communication et technologies*, ed. A. Gras and P. Musso, 419–431. Paris: PUF.
———. 2007a. *L'ideologia delle reti*. Milano: Apogeo.
———. 2007b. Imaginaire et innovation. In *Fabriquer le future 2*, ed. P. Musso, L. Ponthou, and E. Seulliet, 9–61. Paris: Village Mondial.
———. 2009a. Americanisme et Americanisation: du fordisme à hollywoodisme. *Quaderni* 50: 231–247.
———. 2009b. La barbarie managériale. *Cahiers Européenes de l'Imaginaire* 1: 126–134.
———. 2010. Discours de présentation de la chaire Modélisations des Imaginaires, Innovation et Création. Paris: France-Telecom. http://imaginaires.telecom-paristech.fr/a-propos/. Accessed 2 Feb 2015.
———. 2012a. *L'imaginaire des réseaux. Entretien réalisé par Ariel Kyrou, 26 Nov 2012. Culture mobile – penser la societé du numérique*. www.culturemobile.net, paragraphs 1–13.
———. 2012b. *Networks broaden socio-political action, but cannot replace it. The whole world is watching – Projet curatoriale de la session 21*. Grenoble: École du Magasin – Centre National d'Art Contemporaine.
Musso, P., L. Ponthou, and E. Seulliet (eds). 2007. *Fabriquer le future 2*. Paris: Village Mondial.
Musso, P., S. Coiffier, and J.P. Lucas. 2014. *Innover avec et par les imaginaires*. Paris: Manucius.
Noble, D. 1977. *America by design*. New York: Knopf.
Parente, A (ed). 2004. *Tramas da rede*. Porto Alegre: Sulina.
Parrochia, D. (ed). 2001. *Penser les réseaux*. Seyssel: Champ Vallon.
Portugal, S. 2007. Contributos para uma discussão do conceito de rede na teoria sociológica. Coimbra: Oficina do Centro de Estudos Sociais 271.
Roche, L., and T. Grange. 1999. *Innovation et technologie: creativité, innovation et culture technique*. Paris: Maxima.
Saint-Simon, H. 1859 [1822]. Suite a la brochure: des Bourbons et des Stuarts. In *Oeuvres choisies*, tome II. Bruxelas: Meenen.
Scardigli, V. 1992. *Le sens de la technique*. Paris: PUF.
Sennett, R. 2006. *The culture of the new capitalism*. New Haven: Yale University Press.
Sfez, L. 1988. *La symbolique politique*. Paris: PUF.
———. 1992. *Critique de la communication*. Paris: Seuil.

———. 2002a. *Technique et idéologie, un enjeu de pouvoir*. Paris: Seuil.
———. 2002b. La technique comme fiction. *Revue Européenne des Sciences Sociales* [online version] 40–123 (§ 1–68).
Stiegler, B. 2008. *Économie de l'hypermaterial et psychopouvoir*. Paris: Mille et Une Nuits.
———. 2010. *For a new critique of political economy*. Cambridge: Polity.
Subirats, E. 1986. *Da vanguarda ao pós-moderno*. São Paulo: Nobel.
Van Dijk, J. 1999. *The network society*. Thousand Oaks: Sage.
Vitale, Ch. 2014. *Networklogies*. Alresford: Zero Books.

Chapter 7
Paradise, Panopticon, or Laboratory? A Tale of the Internet in China

Dazhou Wang and Kaixi Wang

Introduction

Since the nineteenth century, with the successive development of railway, telegraphy, electricity and the Internet, the figure of the network has become ubiquitous, and the organization of daily life has become a constant use of networks. In his paper, Pierre Musso reviews in detail the origin and development of *retiology*, which is an ideology of networks with utopian aspirations, whose referent is reduced to the fetishism of technical networks, particularly the Internet and teleinformatics networks. He argues that *retiology* is constantly heralding socio-technical revolutions, and thereby relieves social and political utopias of their heavy burden by transferring it to the technological utopia, which has the advantage of materializing. For him, all that remains today are the ideologies of the network, but these are the decayed remnants of a social utopia and conceptual thought developed in the early nineteenth century by philosopher and sociologist Henri Saint-Simon, which conveys the belief that creating a new technical network amounts to triggering a change in society, in the economic mode of production, or even in civilization.

Indeed, for many of the Saint-Simonians, the network is more than a technical artefact built by engineers: it is the symbol of social transformation, facilitating the transformation from conflict to communion, and the engineer-sociologist should be the leading architect of such a social transformation. Saint-Simonianism looks for an ideal society, in which universal association, democracy, peace and liberty prevail. According to P.J. Proudhon, however, Saint-Simonianism's over-symbolization conceals a more concrete issue: the economic policies of network

---

D. Wang (✉) • K. Wang
School of Humanities and Social Sciences, University of Chinese Academy of Sciences, Beijing, People's Republic of China
e-mail: dzwang@ucas.ac.cn

regulation inherent to their very mode of organization. For Proudhon, the network could bring about a social revolution only under certain political conditions of organization and regulation; the choice of a social system is nested within the internal structure of technical networks, and a centralized network means a centralized society and vice versa. In other words, the Saint-Simonian myth of social transformation achieved "automatically" by the development of a new communication network was reformulated by Proudhon, who saw the very architecture of the technical network as a social "choice".

Such an opposition between Saint-Simonianism and Proudhon, as discussed by Pierre Musso, is stimulating and merits being further explored. According to Proudhon, network technologies could have the potential to produce the Saint-Simonian utopia, in which decentralized universal association and freedom reign, or a dystopia, in which centralized power prevails. Moreover, be it utopia or dystopia, the result is determined largely by political choices rather than autonomously. Such a view is definitely in line with the constructionists, who argue that social choices matter in the development and implementation of network technologies. In contrast, the technological determinist approves of the fact that the network technology will produce social transformation autonomously, be it utopia or dystopia. Such a process is not so much a social process as a basically a pure technological process. In fact, with these two pairs of opposition, utopia vs. dystopia, and autonomous process vs. political choice, four possible outlooks on networks can be differentiated, namely, utopia-autonomous realization, dystopia-autonomous realization, utopia-political choice, and dystopia-political choice.

In view of the above, it is clear that Saint-Simonianism has serious shortcomings in its ignorance of the political process involved in the dual construction of network technologies and society, so that "utopia" has long since become synonymous with "fantasy". Universal association, democracy, and equality, all of these as preached by Saint-Simonianism, are desirable. However, to afford the prospect of utopia is not enough, and to pursue it "scientifically" seems to be the worst option, as has been made clear by the experiences of the Soviet Union and Lenin's famous formula: "Communism equals a Soviet regime plus the electrification of the whole country". Paradoxically, the totalitarian state has stemmed partly from Saint-Simonian thought. In this sense, what we need is a political system and process which can safeguard the freedom of mankind. Otherwise, such a "false" ideology would produce fatal errors.

On this basis a variety of views on the Internet can be made clearer. The Internet has been recognized as the perfect symbol of universal association and is able to change the conditions of human existence. It is argued that the establishment of cyberspace as an unlimited space for informational networks affords unrestricted movement in a pure space that is free of friction, ethereal and virtual. As Negroponte pointed out in his influential book, *Being Digital*, the Internet space guarantees everyone is fully able to express their voices. So, it was said that the Internet, "far from being an institution of control, will on the contrary be an instrument of freedom, promising modern humans the ability to shake off the yoke of bureaucracies" (quoted by Musso, Chap. 2). In his *The Internet Galaxy*, Manuel Castells declares that the Internet "is indeed a technology of freedom" (Castells 2001: 1). He argues that the Internet has extraordinary potential for the expression of civil rights, and communication of human values, and provides a new arena for the development of civil society. Jacques Attali even announces that democracy will be electronic and "the political will disappear" thanks to the network (quoted by Musso, Chap. 2). The Internet is thus considered to be anti-hierarchical "in essence", becomes synonymous with self-regulation and equality. If Proudhon were alive today, he would argue that such a positive effect of the Internet is not an autonomous outcome; rather, it could be perverted by "political centralization and economic monopoly" (Musso, Chap. 2). As the advocate of critical constructivism, Feenberg (2014: 146) asserts that the Internet is "an unfinished technology and a terrain of struggle". As a matter of fact, numerous writers note that the authorities in the real world are trying to structure the order on the Internet, and dystopia seems to be a possible prospect. For example, Zuboff (1988) suggests that the Internet will function as the electronic-information panopticon. Indeed, the government is best positioned to regulate the Internet through controlling its underlying codes and structuring the legal environment in which it operates (Lessig 1999). Winner (2014) reminds us that computer users are in danger of being reduced to the condition of techno-serfs, powerless participants in the Net who find themselves fully subservient to the new lords of the realm, and that laws that supposedly protect the rights and liberties of citizens are regularly and secretly breached when it suits the purposes of the military-security-industrial matrix. So there is a sharp opposition between two kinds of social images of the Internet, paradise (utopia) vs. panopticon (dystopia).

To the authors, the question is not so much how to choose either this or that position as how to think about the complicated co-evolutionary process of the Internet and society; and the experiences of China in development of the Internet provide a unique locus to test the above-mentioned thoughts on the political implications of the Internet. As a newly-rising power in the world, China has been experiencing rapid development of the Internet since the later 1980s. At the end of June 2014, China had 632 million Internet users, and the penetration rate was 46.9 %; 527 million of those users accessed the Internet via cell phones, and the proportion of mobile phone usage in netizens has grown to 83.4 %, in excess of traditional PC usage, which is 80.9 %, for the first time, indicating that the mobile Internet phase has arrived.[1] Being referred to as the "entertainment superhighway", the Internet in China has also served as the first public forum for Chinese citizens to freely exchange their ideas. Against the background of the centralized political system, the Internet has affected and been affected by the society at the same time. So, in this paper, the authors aim to narrate a tale of the Internet in China from the

---

[1]CNNIC. The 34th Statistical Survey Report on the Internet Development in China, July 2014. http://www.cnnic.net.cn/hlwfzyj/hlwxzbg/hlwtjbg/201407/P020140721507223212132.pdf. Accessed 20 Nov 2014.

political perspective, and to argue that there is real politics in the social-technological transformation. It is argued that the more suitable image of the Internet is as a laboratory rather than as a paradise or panopticon. In this regard, setting up a proper political framework to govern the co-evolutionary process of the Internet and society is particularly important. The rest of this paper is organized as follows: the sections "To Set-Up the Internet Platform: 1986~1998", "Towards Political Transparency: 1999~2008", and "The Emergence of the Era of Personal Media: Since 2009" narrate the developmental history, divided into three stages, of the Internet in China from the political perspective; the section "The Logic of Internet Politics with Chinese Characteristics" analyses the mechanism underlining the political effects of the Internet in China. The section "Conclusion" concludes the paper.

To Set-Up the Internet Platform: 1986~1998

Unlike the United States, the Internet in China originated from the civil sphere rather than the military one. It was driven by some forward-looking researchers, whose initial purpose was to connect with international advanced universities or institutes for academic exchanges.

In 1986, cooperating with the University of Karlsruhe in Germany, the Beijing Computer Application Technology Research Institute launched an international networking program, the Chinese Academic Network (CANET). In 1987, CANET formally established the first e-mail node and sent out the very first e-mail on September 14, which read "Across the Great Wall, we can reach every corner in the world", representing the beginning of the Internet era in China. Based on the CANET, the Chinese Academy of Sciences (CAS) initiated the National Computing and Networking Facility of China (NCFC) with a World Bank loan in 1989. In June 1991, China's first international special line with direct access to the Stanford Linear Accelerator Center was built by the Institute of High Energy Physics, CAS. In April 1994, NCFC directly connected with the NSFNET in the United States and realized the full function of the Internet in China. In February 1996, developed on the basis of NCFC, Chinese Academy of Sciences Network (CASNET) changed its name to China Science and Technology Network (CSTNET). From 1993 to 1996, besides CSTNET, three other major national networks, namely CERNET (dominated by the Ministry of Education), ChinaNet (dominated by the former Ministry of Post and Telecommunications), and ChinaGBN (dominated by the former Ministry of Electronic Industry), were established respectively. All of these online access routes are owned by the Chinese government, and private enterprises and individuals can only rent bandwidth from them. While CSTNET and CERNET mainly provide non-profit Internet service for scientific research and education, ChinaNet and ChinaGBN provide business Internet services to the public. By the end of the year 1997, all of these networks made connections with the international Internet. With the establishment of ChinaNet, the first Internet service provider in China and a spinoff of the CAS, Beijing InfoHiway

Information and Communication Company was founded in 1996, marking the very beginning of Internet commercial applications in China. In the following year, ISPs such as Sohu.com, Netease.com, and Sina.com were founded successively, all of which were listed on the Nasdaq Stock Market in 2000.

In the early years after the Chinese Academic Network Program was launched, the Internet was only used by a number of top universities and research institutes on a small scale, in order to achieve transnational or cross-regional cooperation in academic studies. Besides that, another group using the Internet in that period was giant transnational corporations which need quick and direct communication with oversea partners. With these particular groups of users, the main use of Internet technology is in the e-mail system. After several years' application on a small scale, with the spread of the Internet technology to other fields, the Internet has begun to play a distinct role in education, economy, culture and social interaction in China. According to the survey by the China Internet Network Information Center (CNNIC), at the end of the year 1998, China had 2.1 million Internet users, and the penetration rate was nearly 0.2 %. The main purposes of browsing the Internet were as follows (response rate): to query information (95 %), to send and receive email (94 %), to download shared or free software (77 %), to chat with people online (42 %) and games and entertainment (35 %).[2]

Considering the economic significance of information technology, the State Council's Information Work Leader Team was established in 1996 in order to promote the advancement of information technologies (including the Internet). One year later, the Ninth Five-Year Plan for State Informationization and Long-range Objective of the Year 2010 was formulated, which listed the Internet as part of the state information infrastructure, and set the goal of pushing forward national economic informationalization by vigorous development of the Internet industry. In 1998, the Ministry of Information Industry (restructured in 2008 as the Ministry of Industry and Information Technology), by merger of the former Ministry of Posts and Telecommunications, the former Ministry of Electronic Industry and the networks regulation department of the former State Ministry of Radio, Film and Television, was established to promote the development of the information industry, maintain fair market competition and take charge of examining, authorizing and issuing of Internet Service Providers' operating licenses.

At this stage, almost all the practices the Party and government took were in response to those threats incubated from the international environment. Although the government set few restrictions on the applications of the Internet, and the public interconnected network was completed as early as 1996, it did not popularize quickly because the economy was still underdeveloped and the computer was still unusual in people's daily lives. Even so, Chinese scholars at that time put their hopes in the development of the Internet, not only because it belongs to the "first

---

[2]CNNIC. The 3rd Statistical Survey Report on the Internet Development in China, January 1, 1999. http://www.cnnic.net.cn/hlwfzyj/hlwxzbg/200905/P020120709345373005822.pdf. Accessed 10 Nov 2014.

productive force", but also because it has the potential to lead Chinese society to a higher level of freedom and equality. For example, Liu Ji, the former vice president of the Chinese Academy of Social Sciences, and his colleague Jin Wulun once wrote in their book on the knowledge economy in the very stages Internet implementation in China: "The Internet is such an independent space, in which anyone in any place can express her views freely. No matter how bizarre her opinion is, she does not need to worry about the fact that she could be forced to remain silent or be suppressed to be obedient" (Liu and Jin 1998: 278).

The year 1999 witnessed a milestone in the history of the Internet politics in China. On April 25, more than 11000 FalunGong believers conducted a sit-in demonstration at the Tiananmen Square and even surrounded Zhongnanhai, the location of the state leaders' office, for 8 h, requiring the central government to legalize the FalunGong organization, and to release 45 followers who had been arrested by the Tianjin Public Security Bureau a week earlier. Considering that FalunGong is not a loose organization but a "tightly-organized" and "secretly-structured" one based on modern information technology, i.e., the Internet and the mobile phone, the Central Committee of the Communist Party of China (CCCPC) announced it was an "evil organization" 3 months later, and thereafter its core members were arrested. This seemed to be the first time that the Internet manifested its political implications in China. From then on, fighting the FalunGong has become one of the major goals of the government's regulation of the Internet.

Towards Political Transparency: 1999~2008

Another landmark was the establishment of the BBS of People's Daily almost at the same time. On May 8, 1999, the United States bombed the Chinese embassy in Yugoslavia. This provoked a burst of sentiment among the public, reflecting a pent-up need for free expression. In the afternoon of the next day, the People's Daily set up a platform called Strong Protest against the NATO Ferocity BBS. Later, on June 9, its name was changed to the BBS of the People's Daily, which accidentally opened up a communication channel between the public and the government, and acted as the most influential platform ever for the expression of public opinion.[3] Of all posts there, about one third is criticism, one fourth is suggestions and only one tenth is praise. Every year since 2001, this BBS organizes a poll for the "Annual Top-Ten Influentials", who had been actively digging for problems and contributing ideas to solve them on behalf of all concerned netizens in that year. The political potential of these voices on the Internet became much clearer. It was in 1999 that the State Council launched the "government online project", aiming to establish

---

[3]This was possibly a clever use of the BBS to shift the attention of the public from the 10th anniversary of the June 4 Incident. And demonstrations by Falungong believers before this time could be perceived by the authorities as a provocation.

government websites so as to promote government department office automation and the quality of public services. Later on, at the end of the same year, the "Enterprise Informationalization Project" was launched by the former State Economic and Trade Commission.

From 1999 to 2008, China had experienced a rapid development of the Internet, thanks to the significant effort the government had made. In 2002, the Specialized Plan for Informationization in the Tenth Five-Year Plan for National Economic and Social Development was promulgated. This defined China's priorities in this regard, including promotion of e-government, vigorous development of the software industry, strengthening of development and utilization of information resources, and acceleration of the development of e-commerce. The State Informationization Strategy (2006–2020) is also worth mentioning. It was formulated in November 2005, and clarified the priorities of Internet development as promoting national economic informationization while adjusting the economic structure and transforming the patterns of economic growth; building e-government while enhancing the capability of governance; and spurring the informationization of social services while building a harmonious society. In this context, the nearly simultaneous use of Web 1.0 and Web 2.0 contributed to the striking surge in the population of netizens in China (Fig. 7.1). As living standards, especially of urban residents, improved dramatically during the first decade of 21st century in China, people were motivated to seek more information and social connections in their daily lives. Such applications of Web 2.0 as instant messaging tools, BBS and SNS, have made relationship-building and maintenance much more convenient and direct than in real life. As the first SNS website in China, Renren.com was founded in December 2005, with college and high school students as its major users. Then Kaixin.com was founded in 2008, with its main users being white collar groups in the city. Both of them focused on entertainment and communication within circles of acquaintances and required real-name registration. Basically, such applications functioned in their own way, and lack of connections among them greatly hindered the realization of universal networking on the Internet.

Along with the development of the Internet, the scale of people's participation and the scope of political affairs has gradually increased. Different kinds of voices, expressing dissatisfaction, appeals, requirements, suggestions, even anger, were all reflected online and formulated an ever-growing countervailing power. As a result, a brand-new social group called e-influentials has arisen online. They can create or manipulate public opinion by posting motivational messages, articles or videos on platforms such as chatrooms, portal websites, BBS and blogs. The e-influentials can be divided into two categories: the initiative Internet media and perspicacious individuals. The former were more profit-oriented. Facing competition from the traditional or the government-led offline media, these small and private media have advantages in terms of flexibility and swiftness. Posting something new or different can help them to draw attention from the public, which can turn into concrete profits. These media are greatly motivated to guide public attitudes and to act as the public's voice on political issues at times. In contrast, individual e-influentials, most

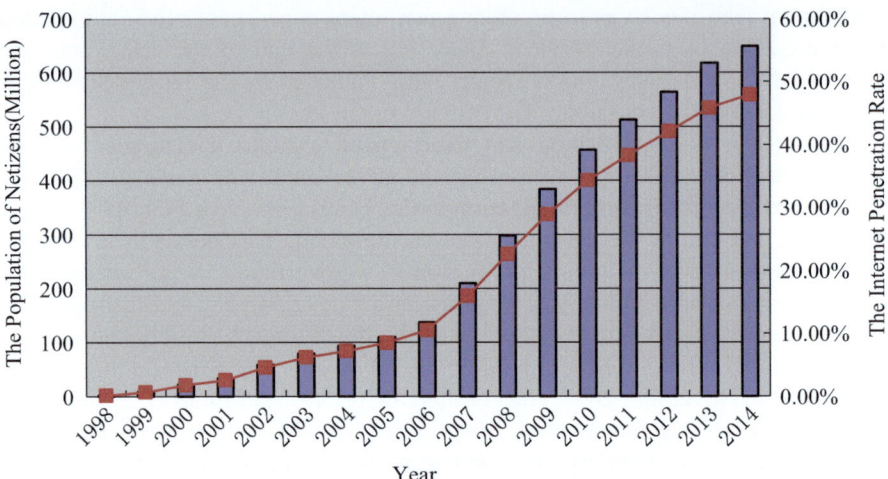

Fig. 7.1 The population of netizens and the Internet penetration rate in China since 1998 (Source: China Internet Network Information Center (CNNIC). The Statistical Survey Report on the Internet Development in China (since 1998), http://www.cnnic.net.cn/hlwfzyj/hlwxzbg/. Accessed 1 Feb 2015)

of whom are public intellectuals or celebrities, usually do not speak for profit. Instead, they speak for what they are interested in and want to make a difference by posting their opinions on personal web pages, blogs or BBS. They either expose problems existing in social life around them, appealing for attention to particularly vulnerable groups, or simply express their anger or dissatisfaction. Either way, they enjoy discourse rights empowered by the Internet which could not be achieved under traditional social conditions.

This situation was clearly demonstrated by the Sun Zhigang case. On March 17, 2003, a young university graduate named Sun Zhigang from Hubei Province was taken to the hospice in Guangzhou by the police, as he did not have the required Temporary Residence Permit. On March 19, he was relentlessly beaten by staff there and the next day, they announced his death from a heart attack. At first, a journalist from a local newspaper, the Southern Metropolis Daily (SMD)[4] obtained brief information from a college student's SMS message. With the editor's approval to conduct an interview, the journalist contacted and helped Sun's father to engage a lawyer and ask for a forensic autopsy in order to obtain legal evidence. On April 18, 2003, autopsy results proved that he had been beaten up before he died,

---

[4]The Southern Metropolis Daily started on January 1, 1997, led by the deputy editor Cheng Yizhong, who believes that this newspaper can execute the traditional duty of the intellectuals by following the model of the Washington Post to play the supervision role in preventing the abuse of power. In fact, for thousands years, a sense of responsibility has motivated Confucian scholar-bureaucrats to devote themselves to societal development. As an old Chinese saying goes, "Every man alive has a duty to his country."

resulting in widespread soft tissue damage, and ultimately his death. In order to prevent possible interference from the local government's propaganda department, the SMD edited the news release overnight and reported it with a splash headline, "The Death of the Detainee Sun Zhigang", after its deputy editor had emailed it to the editors of Sina and Sohu, two major ISPs in China. Both set up specific discussion zones or message boards for this incident. It was just at a time when China was combating the new epidemic disease SARS, i.e., atypical pneumonia, and temporarily eased its strict limits on media reporting under great pressure both at home and abroad. Through setting up special news topics, a main page title link, a message board, BBS, and conducting online survey and online guest interviews, major web portals all played their distinct roles. As a result, Sun Zhigang became famous overnight, and his story occupied the main pages of nearly all the media in the following days. On May 4, the People's Daily Online published an article entitled "Sun Zhigang Case, Who Pretends to be Dumb?" which drew the attention of officials at all levels. Under such pressure, officials from the central government put forward clear requirements that the facts should be investigated and the criminals punished. On May 13, all of the responsible personnel were caught and convicted. On June 20, the State Council formally abolished the Regulation for Custody and Deportation of Urban Vagrants which was issued in 1982,[5] and at the same time promulgated Order No.381, the Regulation Measures for Urban Vagrants' Assistance.

In any event, the public has generally been empowered to access information on almost all kinds of social affairs and has enjoyed knowledge and disclosure rights for the first time. In this way, Chinese politics started to became somewhat transparent. The government was aware of the great power of public opinion online and of collective actions through the Internet, and at times was willing to adopt an enlightened strategy to deal with them. For example, Article 15 of the Regulations of the People's Republic of China on the Disclosure of Government Information, which was promulgated and put into force in 2008, stipulates, "Government agencies should take the initiative to disclose government information, which should be disclosed by means of government gazettes, government websites and press conferences, as well as through newspapers and magazines, radio, television and other methods that make it convenient for the public to be informed." The central government required governments at all levels to establish corresponding mechanisms and give prompt explanations on issues of public concern.

The government, however, fearing social instability, especially considering that the summer Olympic Games were scheduled to be held in Beijing, determined to exercise stronger regulation of the Internet. This took place in two ways: one was the regulation of online information or service providers; the other was the

---

[5]Custody and repatriation was an administrative procedure, established in 1982 and ended in 2003, by which the police in the People's Republic of China (usually in cities) could detain people if they did not have a residence permit or temporary living permit, and return them to the place where they could legally live or work (usually rural areas). At times the requirement included possession of a valid national identity card.

regulation of the content of disseminated information. For those information or service providers, enterprises or organizations which registered at the Industry and Commerce Administrative Department, the government applied the licensing and filing system (Chen 2010). Under such a system, to set up a website, the essential condition is to get a licence from the Internet regulation agencies. All information which it is planned to put on the website has to be filed and recorded by the agency for further reference before it is published online.[6] As for those random information distributors who are difficult to trace, mostly individuals, the early prevention practice is to formulate a real-name registration system. According to Article 23 of the Regulations for the Administration of Business Sites of Internet Access Services issued on 14 August 2002, all business sites providing Internet access services must check and register the valid certifications as ID cards of their customers, and keep their web browser records. Subsequently, the Provisions on Further Strengthening the Management of the Campus Networks of Colleges and Universities, issued by the Ministry of Education in 2004, clearly requires applying real-name systems on those campus networks of colleges and universities, including internal communication platforms as BBS. On November 29, 2004, the Internet Trust and Self-discipline Alliance, co-established by the Internet companies Sina, Sohu and Netease, proclaimed self-disciplinary regulations for China's Internet wireless service providers (SPs), representing the continued and serious efforts to implement self-discipline of the wireless Internet SPs in China. Moreover, the government started to use technical means to trace the IP addresses of those issuing "false" or "harmful" information and to apply administrative punishments to the personnel involved.

In fact, China has built up its so-called "Great Firewall", an expression coined in an article in *Wired Magazine* in 1997, to block access to some particular websites or homepages which are alleged to contain unhealthy information or to international websites regarded as threats to national security (Wang and Hong 2010; Kim and Douai 2012). As the main part of the Great Firewall, the Golden Shield Project started in 1994, aiming to build a massive surveillance and censoring system. The first part of the project lasted 8 years and was completed in 2006. The second part began in 2006 and finished in 2008. The main operating activities of the gatekeepers at the Golden Shield Project include monitoring domestic websites and email and searching for politically sensitive language and even calls to protest. Another more specific measure is the filtration of sensitive words, mostly names of government or party leaders or public departments, and words relating to major controversial public issues, and some rude words. Its application ensures that messages or videos containing those words will be blocked from the public. Its coverage extends to almost all current online platforms, Internet media and their portals, search

---

[6]For instance, at the portal website of Sina, there are more than ten types of license listed at the bottom of its homepage, among which the top three are: the license to post news reports online, the license to make their own videos or shows, and the commitment to avoid illegal or harmful information.

engines, private homepages, BBS, instant messaging tools and e-mails. Since 2007, seven laws or regulations on the Internet have been issued, changing the regulator from the State Council and the Ministry of Information Industry to those more specific functional agencies, representing political moves to decentralize regulation of the Internet (Chen 2010).

Faced with these strict regulations, however, netizens have come up with a batch of countermeasures to break the restrictions formulated by the government, such as firewall bypassing, which refers to the behavior of Internet users to resort to software or a proxy to get access to the websites or online resources that are blocked by the Great Firewall, mostly for information, social networking, and entertainment (Yang and Liu 2014). Given the large amount of information spreading online, manual monitoring is limited in its capabilities, so word filtering has been implemented instead through particular settings mainly embedded in router, application server and terminal software.[7] Resourceful netizens have figured out various transformative expressions to replace the sensitive words, as illustrated in Table 7.1. Once having agreed on such invisible codes, people can continue to communicate and spread the information that is supposed to be blocked.

It is worth emphasizing that the year 2008 was special. At the beginning of that year, the international financial crisis was spreading from the US to the whole world. The bull market in China was coming to an end, and this was followed by the Wenchuan Earthquake on May 12 and the Weng'an Riot in Guizhou Province on June 28.[8] In August, Beijing held the 2008 Summer Olympic Games, which was praised as the exceptionally successful event. In September, however, the melamine scandal broke out, giving rise to fierce discussion on the Internet. By the end of 2008, the total number of Internet users in China reached 298 million, jumping to first in the world for the first time; the penetration rate reached 22.6 %, surpassing the global average. In short, in 2008 both the economy and politics in China appeared to reach their peak and then began to decline, while the regulation of the Internet seemed to be more and more effective, and Internet applications started to evolve into a new stage.

---

[7]Search results for sensitive terms in Chinese search engines are filtered. These Chinese search engines include both international (e.g., yahoo.com.cn, Bing, and formerly Google China) and domestic ones (e.g. 360 Search and Baidu). Attempting to search for censored keywords in these Chinese search engines will yield few or no results, or display the following: "According to the local laws, regulations and policies, part of the search result is not shown".

[8]The Weng'an Riot took place on June 28, 2008, involving tens of thousands of residents in Weng'an County, Guizhou Province. Rioters smashed government buildings and torched several police cars to protest against an alleged police cover-up of a girl's death. See, http://en.wikipedia.org/wiki/2008_Weng%27an_riot. Accessed 16 Oct 2014.

Table 7.1 Some of the sensitive words and their transformations

Sensitive words (A)	Transformation (B)	Correspondence of A and B
Harmony	River crab	Same pronunciation in Chinese
Party (in Politics)	Party (for fun)	Same spelling in English
June 4	March 35	Same meaning
Autarchy	A pair of scissors	B is the literal meaning of A
State leaders	Optimus Prime	Metaphor from the movies
Tank	Chariot	B is the general reference of A

Source: Collected from several popular websites

The Emergence of the Era of Personal Media: Since 2009

Since 2009, with the emergence of Weibo, a kind of Chinese microblog, and the increasingly wide use of smart mobile phones, China has gradually stepped into the era of personal media. In 2009, the number of Chinese Internet users reached 384 million, and the penetration rate reached 28.9 %; the number of mobile Internet users reached 233 million, with annual growth of 106.0 %,[9] as third generation (3G) licenses began to be issued to mobile service suppliers in January 2009. The successive years witnessed explosive growth in the number of both Weibo and mobile Internet users: by the end of September 2014, Sina Weibo's monthly active users reached 167 million,[10] and by the end of June 2014, mobile Internet users reached 527 million. With the development of the mobile Internet and the emergence of the personal media era, the political implications of the Internet have become plainly manifest in China.

The start of this stage, the year 2009, was sensitive, for it was the twentieth anniversary of the June 4 Incident. The government accordingly sought to exercise strict controls over the Internet. On May 19, 2009, the Ministry of Industry and Information Technology released a notice that required all computers in production and sales in China to pre-install the "Green Dam-Youth Escort" Internet filtering software. It was said that the installation was intended to block violent and pornographic content on the Internet, and to protect children. Such a move, suspected to be a new monitoring measure, aroused huge opposition, and the minister later clarified that installation was compulsory only for computers in

---

[9]CNNIC. The 25th Statistical Survey Report on the Internet Development in China, January 2010. http://www.cnnic.net.cn/hlwfzyj/hlwxzbg/201001/P020120709345300487558.pdf. Accessed 14 Nov 2014.

[10]Weibo Data Center, Report on Development of Weibo in the Year 2014, http://data.weibo.com/report. Accessed 20 Jan 2015.

public places, including schools and Internet cafes, and individuals could install the filtering software on their computers voluntarily. Later on, however, the requirement was postponed and in the end came to nothing. On May 10, 2009, the attention of almost the whole county was drawn to the Deng Yujiao Incident,[11] which took place in Hubei Province, and the interaction between netizens and the governments by and large followed the mode of the Sun ZhiGang Incident. On June 26, 2009, the Ministry of Culture and Ministry of Commerce jointly issued its "Information on the Management Work of Virtual Currency Trading for Online Games", to standardize the trading behavior of suppliers and participants in online games. This regulation provided new technical tools for controlling information content besides the filtering of sensitive words: post deletion and "transmission forbidden", which are more manual and comprehensive. Information gatekeepers are everywhere, searching and erasing "irrational" or "false" discussions. For some serious situations, posters or transmitters of that information can be investigated and held accountable.

It was in this year that Facebook and Twitter were blocked by the Chinese government following deadly riots in the Xinjiang Autonomous Region,[12] which were believed to be abetted by social networking sites. For the same reason, Fanfou, an imitator of Twitter and a predecessor of the microblog site in China, was shut down in July. In fact, when the riots began, communications were immediately cut off. At that time, Fanfou became almost the only information channel for ordinary people to contact their relatives in the region, and its media feature was surprisingly amplified. It exceeded traditional media and web portals in terms of diversity and speed of communication, which was also the fatal defect in China because Fanfou had no capability of dealing with the dissemination of sensitive or false information, being regarded by the government as the responsibility of the Internet companies. In the very early days of the development of the microblogging service, the tolerance of the government for this new media was relatively low, and the government also had no specific measures to filter this kind of SNS, so shutting down the services at that moment seemed to be the only choice.

---

[11] Deng Yujiao, a 21-year-old pedicure worker, tried to rebuff the advances of Deng Guida, director of the local township business promotions office, who had come to the hotel seeking sexual services. She allegedly stabbed her assailant several times trying to fight him off, resulting in his death. Badong County police subsequently arrested Deng, charged her with homicide, and refused to grant her bail. This case came to national prominence through Internet forums and chatrooms, where netizens were enraged by her treatment. Following strong public protests and online petitions, prosecutors dropped the murder charges, and she was found guilty but did not receive a sentence due to her mental state. The two surviving officials involved in the incident were sacked in response to public pressure. See, http://en.wikipedia.org/wiki/Deng_Yujiao_incident. Accessed 11 Oct 2014.

[12] The July 2009 Ürümqi riots were a series of violent riots over several days that broke out on 5 July 2009 in Ürümqi, the capital city of the Xinjiang Uyghur Autonomous Region. China officials said that a total of 197 people died, with 1721 others injured and many vehicles and buildings destroyed.

Sina clearly understood the root of Fanfou's failure, and was also clear why Twitter was banned in China: they were unable to solve the problem of monitoring the release of "sensitive" or "false" information. To Sina, the key lay in a set of appropriate censoring mechanisms. Such mechanisms not only had to comply with the regulations on information safety, but also make users feel comfortable rather than weary or hostile.[13] In contrast to Fanfou, Sina, with so many years of operation of releasing news, BBS and blogging, has acquired a unique ability to handle such problems and to win the trust of regulators. In September 2009, Sina thus launched the Sina Weibo service, just 2 months after Fanfou had been shut down. Very quickly, Sina Weibo achieved great success by making full use of celebrity resources and developing a mechanism for censoring sensitive information.[14] So, just 1 year later, at the end of September 2010, users of Sina Weibo reached 30 million. Other portals such as Sohu, Netease, and Tencent followed up successively to launch or re-launch their own microblogging services in 2010. With the explosive growth of users, the tremendous role of the microblog as the personal media in the political life was manifested in the year 2010. It is in this year that the "microblogger" was awarded the Person of the Year by the Southern People Weekly, with comments as follows: "Something taboo in the traditional media was discussed in the microblogging space. Some said this, and others said that. As a result, the sensitive things gradually desensitized and became the mundane, and the space for freedom of speech expanded little by little".[15]

The Yihuang Self-immolation Incident, among others, demonstrated clearly how an originally local incident could be transformed into a high-profile public event with the live show on the Weibo. The Incident occurred on 10th September 2010 in Yihuang County, Fuzhou City, Jiangxi Province.[16] On that day, local government authorities with demolition workers came to the home of the Zhong family. Ye Zhongcheng and two other family members climbed to the roof and burnt themselves in protest, and Ye later died. The Zhong family lived in a three-storey house, which was owned under three licenses belonging to three Zhong brothers. In 2007, the home of the Zhongs and those of 21 neighbors were to be removed for the building of a new bus station. The Zhongs were given two options: monetary compensation or another house some sixty meters away. The Zhongs refused both options and responded with other demands. Negotiations were long drawn out and finally failed. On April 18, 2010, the family's electricity was cut off, so they purchased a petrol-powered generator. At about 9 a.m. on the 10th September 2010, over 40 police and workers from the Chengguan City Urban Administrative and Law Enforcement Bureau and government officials arrived at

---

[13]Rong, Zhenhuan, Recall: why the originators of microblog in China became martyrs? http://www.cyzone.cn/a/20130809/244335.html, August 9, 2013. Accessed 9 Oct 2014.

[14]Sina Weibo was listed on Nasdaq on April 17, 2014.

[15]Southern People Weekly, 2010, No.45.

[16]See, Southern People Weekly, 2010, No.45, and Yihuang self-immolation Incident, in http://zh.wikipedia.org/wiki. Accessed 29 Sep 2014.

the Zhongs' house. They said they needed to enter the house to check the gasoline supply. Being asked for a search warrant, the officials said they didn't need one as it was an emergency. The Zhongs locked their door. Several minutes later, the police broke the door down and entered the building. Zhong's wife Luo Zhifeng, daughter Zhong Ruqin and his brother Ye Zhongcheng went to the roof and burnt themselves with gasoline. On September 16, two sisters of the Zhongs were going to Beijing to attend a program recording at the invitation of the Social Visibility Column of the Phoenix Satellite Television, which meant the case would be publicized widely. However, they were believed to be going to Beijing to petition and were thus blocked by a local authority official. They had no choice but to hide in the female toilet cubicle at the Nanchang Airport, and made a phone call for help to Liu Chang, a reporter with the New Century Weekly who had interviewed them previously. Liu Chang posted his first weibo about the situation in half an hour, appealing for public attention. Twenty minutes later, one of the network influentials, Murong Xuecun, forwarded this massage on his weibo account. From then on, forwarding and discussion of this post on Sina Weibo began to increase geometrically. On that morning, the original post was forwarded more than 2700 times and with more than 1000 comments. 50 min later, Deng Fei, a journalist with the Phoenix Weekly Magazine, got the news by QQ instant massaging from Liu Chang, and started a live show on his weibo space about the struggle between the Zhong sisters and officials of the local authority. Later, more and more netizens on Weibo gathered and carried out the rescue and love relay. Qiu Jianguo, the county Party secretary, began to receive phone "greetings" (threatening messages from the netizens). Newspaper reporters rushed to the scene and started to report synchronously. As a result, the Zhong sisters were able to regain their freedom and agreed upon a proposal to negotiate with an upper-level government official in the presence of the media next day. Under the advice of a reporter, the Zhong sisters registered a weibo account and began to post messages on it at night. To this end, Zhong Rujiu got a smartphone in order to tweet better. Previously, they had only played the "stealing vegetable" game on the Internet through their mobile phones. Overnight, the number of "fans" focused on their weibo approached 20,000, and each of their posts was commented on and forwarded by tens of thousands of netizens. On September 26, the condition of the badly burned Luo Zhifeng and Zhong Ruqin was deteriorating, so the Zhong sisters and their lawyer started to seek medical help on Weibo. Seeing this, Beijing Chuizi, a microblogger with numerous fans, finally decided to do something for the Zhongs and posted his first weibo for help. Previously, a tragic picture of Zhong Ruqin's burning into a ball of fire while falling from the second floor had drawn his attention. A netizen responded at once that the Beijing No. 304 Hospital of the People's Liberation Army had the best burns surgery, and at the same time gave a phone number to Beijing Chuizi. He dialed and to his surprise, the rescue request was allowed. He then contacted the Beijing Red Cross for a special emergency plane to take the wounded to Beijing. In the meantime, another group of netizens had contacted medical experts in Hunan and Guangdong Provinces respectively. However, in view of condition of the wounded at that time, the more feasible way was to invite the experts to Nanchang rather than escort the wounded to Beijing or elsewhere. Beijing Chuzi posted another message on Weibo, telling netizens that the experts from the No.304 Hospital had agreed to go to Nanchang, and asking for their help in handling local hospital procedures. Later on, the netizen Dong Chongfei got the phone number of the director of the No.1 Hospital affiliated with Nanchang University, in which the three wounded had been treated. Thankfully, the director's attitude was very open and willing to arrange co-operative treatment. It was then 3 a.m. On the same day, Chai Jiake, director of burns surgery of the Beijing No.304 Hospital, left for Nanchang for medical consultation. A few days later, the two wounded were transferred to Beijing No.304 Hospital for follow-up treatment, and they eventually got out of danger. Finally, with the help of Weibo, the incident drew the attention of the central leadership, who instructed the respective agencies to conduct a seriously comprehensive investigation. As a result, both the Party secretary and the head of the County were dismissed, and other officials were also punished for abuse of power.[17]

Through the personal media, Weibo, the Yihuang Incident was successfully made public and settled through collective action, accompanied by extensive discussions of the legitimacy of the forced demolition and the reform of land system, which would have to take years to be resolved. Anyway, it seems that the government and the Internet service providers have reached a tacit agreement, by which the latter voluntarily implemented self-censorship in relation to content released. However, Google was a notable exception. Since launching "Google.cn" in 2006, Google had been reluctant to cooperate with Chinese government in filtering the Internet and was thus punished for containing pornographic contents in its search results. In March 23, 2010, Google announced its withdrawal from the Chinese market, criticized the censorship system, and appealed for Internet freedom, drawing global attention to the issue of the legitimacy of the Chinese Internet regulation system (Kim and Douai 2012). On June 8, 2010, the State Council Information Office issued its white paper on China's Internet status for the first time, claiming that the Chinese regulations "clearly prohibit the spread of information that subverts state power, undermines national unity, infringes on national honor and interest, incites ethnic hatred and secession, advocates heresy, pornography, violence, terror, and other information that infringes upon the legitimate rights and interests of others". According to these regulations, basic telecommunication business operators and Internet information service providers shall establish Internet security management systems and utilize technical measures to prevent the transmission of all types of illegal information".[18] On the other hand, the

---

[17]Ironically, under the existing administrative regulations on urban demolitions, local governments do have the right to adjudicate, apply, and perform compulsory demolition and removal, which would likely result in "legitimate" damage to the citizens' houses, even personal injury. This indicates that China needs to amend the relevant regulations as well as reform the land system.

[18]Information Office of the State Council of the People's Republic of China. The Internet in China (White Paper), June, 2010, Beijing.

government has also tried to reach out to listen to the public's voices. For example, on September 8, 2010, the People's Daily Online and the News Network of the CPC officially launched the "Message Board with Straight Connection to Zhongnanhai", which was designed to allow netizens to express their opinion and suggestions to the central leaders.

In this context, Weibo had surely witnessed a period of golden times. According to a survey by CNNIC, Weibo users in China reached 250 million in 2011, with growth of 296.0 % over the previous year. As the Sina Weibo Data Center recorded, by the end of the year 2012, the number of registered users on Sina Weibo reached 500 million, with 46.2 million daily active users. Compared to such online communities as BBS, Blog, and SNS, Weibo services are more diverse and dynamic, and offer more opportunities for netizens to participate in political life.[19] On Weibo, more and more people have participated in discussing sensitive affairs, and serious discussions along with high public attention have always generated a strength of public opinion which can force the government to resolve issues in the right way (Wang et al. 2013; Xiong et al. 2013). This situation came to a climax when the Guo Meimei Incident[20] erupted on 21st June 2011 and with the Wenzhou High-speed Train Collision Accident[21] on the 23rd July 2011. The former raised doubts about state-owned charity organizations, and the latter highlighted defects in the system of Chinese news reports and the handling of accidents. It seemed that those discussions on Weibo were exceeding the government's tolerance limits.

---

[19]Certainly, traditional SNS also has its political implications. Mou et al. (2013) reveal a moderate but positive impact of online social networks on online civic participation, lending qualified credence to the positive role of social media in facilitating recent events such as the Arab Spring and other grassroots protests, and suggests that online social networks and online political discussion are positively related.

[20]Guo Meimei forged her identity on Sina Weibo by proclaiming herself a manager of the Red Cross Society of China. Her display of pictures showing a lavish lifestyle driving Mercedes and owning a big mansion drew skepticism from people who questioned if her wealth came from the Red Cross Society of China and if the charity had misused its donation from the public.

[21]At 20:30 on July 23, 2011, two bullet trains collided and caused the death of 40 passengers. 13 minutes later, a Sino Weibo user named Yangquanquanyang posted a message for help, saying that the bullet train D301 had derailed near the South Station of Wenzhou, kids all crying in the carriage and no one was coming for rescue. Several other passengers on site also managed to send out messages on Weibo. The traditional and Internet media were left far behind, becoming the passive receivers of that first-hand information and having to interpret the whole accident from messages on Weibo, piece by piece. In the initial stages after it had taken place, and because of the delays and obfuscation surrounding the causes of the accident and the rescue measures implemented by the Ministry of Railways, as well as the inappropriate remarks of officials, saying "believe it or not, anyway, I believe!", the whole Weibo space witnessed a heated debate on the functional and behavioral anomalies of government officials. Subsequently, as negative voices became dominant online, the traditional media started to change their attitude from covering up to questioning. Finally, with a general appeal for an impartial investigation, on July 27th, premier Wen Jiabao hosted the No.165 executive meeting of the State Council, obtaining debriefs on this major traffic accident from different departments, requiring them to post their investigating and handling process publicly, and to organize and publish the final investigation report on a timely basis.

Fearing the increasingly heated social interaction online, on 16 December 2011, Beijing took the lead in requiring municipal microblogging service providers to register users with real names. The rules also required websites to obtain approval from the Beijing Internet Information Office (BIIO) to operate microblogging services in Beijing, and to ensure the authenticity of their users' identities. The regulations were to be fully implemented within 3 months. Later on, Guangzhou, Shenzhen and Shanghai followed suit. As the registrations are required at the backstage, microbloggers are free to choose their screen names, and users who are only browsing don't need to submit their real names. Just in that period of time, the Wang Lijun Incident occurred in February 2012, triggering another Internet carnival. Amidst rumors of political infighting with Chongqing Communist Party Chief Bo Xilai, Wang Lijun, then vice-mayor of Chongqing, arranged a meeting on February 6 at the U.S. consulate in Chengdu, where he remained for approximately 24 h. He left the consulate of his own volition and was taken to Beijing. The incident led to the abrupt end of the political career of Bo, who was seen as a contender for a top leadership position at the 18th Party Congress in 2012. The Bo Xilai Case and subsequently the Zhou Yongkang Case[22] then became the center of the netizens' attention, and Weibo functioned as the platform for the political infighting between their supporters and opponents for a period of time.

Indeed, Weibo has manifested its huge strength in Chinese political life. With Weibo, the public are no longer confined to discussing social issues, but also attempt to find solutions to problems with their combined strength, mediated by the Internet. And in most cases, such attempts develop into collective actions. In this context, even a piece of posted message or picture can produce strong collective action and move the country forward.[23] Most of the time, the Chinese government seems to allow netizens to appeal for corruption and abuses of power to be

---

[22]On July 29, 2014, the Communist Party's Central Committee announced a probe into former security chief Zhou Yongkang over alleged discipline violations, confirming the rumors about him spread online for a period of about 2 years. Zhou was a member of the Politburo Standing Committee, the party's top decision-making body, from 2007 to November 2012. He also headed the Central Politics and Law Committee, the party's top authority on domestic security affairs. It was said that Zhou Yongkang was closely involved with the Bo Xilai Case.

[23]At 14:22 of January 17th, 2011, Yu Jianrong, a professor from Chinese Academy of Social Sciences posted a picture of a young boy who had been abducted in 2009 and forced to beg for money on street. The picture was taken by a netizen during the Spring Festival of 2010, and the helpless expression on his face was in extreme contrast to the jubilant atmosphere of New Year. This message dropped like a bomb among the public and led to extensive discussion and further transmission. On January 24th, 2011, Yu Jianrong, uniting with several celebrities, reporters and authors, launched a charity activity called Taking Photos to Save Child Beggars on Weibo, appealing to people to take pictures of the children begging on street and to post them online with their locations, hoping the parents who had lost their children might get the message. Later, this activity expanded to capturing human traffickers under the joint efforts of the police, media and civil volunteers; creating a special fund for child beggars under the One Foundation to provide financial support for the activity; More importantly, several CPPCC Members, like Han Hong and Liu Hongyu, submitted a proposal to save child beggars legally, through the National People's Congress. With the united efforts from different social forces, the related legal provisions are around the corner.

checked, but only at the local level. If the action is directed to the central leadership or conceived as a subversion of CPC or state power, then it will be defeated. It seemed that the heated discussions on the Bo Xilai and Zhou Yongkang cases would go that far, due to their high position in the party, and would pose potential threats to the central leadership and the political regime. Therefore, once the dust had settled, tightening the rules on free speech seemed to be the only choice. Unsurprisingly, in mid-2013, the Chinese authorities sought to reassert strict rules on public speaking, with an emphasis on the conduct of e-influentials, so-called opinion leaders, online. For example, two arrested users were said to have "incited dissatisfaction with the government" by spreading rumors about Lei Feng, a deceased soldier who is often taken as an example of the model Chinese citizen. They confessed the crimes to the police and admitted that they had said the Internet users should be manipulated into believing that they are the "victims of social injustice" and only "anti-social activities could help them vent their dissatisfaction".[24] Local governments responded rapidly by arresting more local opinion leaders, which raised concerns over free speech in cyberspace. Moreover, such actions contributed directly to lower activity rates and fewer highly active accounts on Weibo, since the crackdown on rumors has made the passing along of news (even news that has nothing to do with politics) dangerous unless it comes from an official source, and this in turn has made Weibo more boring.[25] At its height, Sina Weibo had more than 600 million registered users. But the microblogging activity had been dropping off since late 2011, and dropped precipitously in the fall of 2013. Fortunately, since the end of August, the direction of the wind of official propaganda has changed a little bit. The authority issued a document to require that the crackdown on rumors on the Internet shall be in accordance with the rule of law, and shall strictly prevent escalation of the action.[26] On September 2, Xinhua News Agency issued a commentary, saying that "the initial intention to crackdown on rumors on the Internet was correct, but nobody has the right to put a label of rumor on any voice he dislikes, and vigilance shall be maintained in order to avoid the abuse of power and off-tracking". On September 4, the People's Daily also published a commentary entitled "To grasp firmly the Internet as the biggest variable", saying that "to contain the vitality of the Internet was also contrary to the intention of CCCPC and the trend of the times... We should be good at transforming the vitality of the Internet into dynamics of social progress, rather than letting it be alienated as resistance and destructive power".[27] To help with legalization and standardization of the punishment mechanism beyond post

---

[24]Two arrested in Web rumor crackdown, *Global Times*, 2013-08-21. http://english.sina.com/china/2013/0821/621142.html. Accessed 12 Oct 2014.

[25]Custer, Charles, The Demise Of Sina Weibo: Censorship Or Evolution? Forbes Asia, 2/04/2014. http://www.forbes.com/sites/ccuster/2014/02/04/the-demise-of-sina-weibo-censorship-or-evolution/. Accessed 13 Nov 2014.

[26]Liu, Jun, and Jing Ju, Behind the scenes of "the crackdown on Internet rumors", *Southern Weekend*, September 5, 2013.

[27]To grasp firmly the Internet as the biggest variables. *People's Daily*, September 4, 2013. http://news.xinhuanet.com/comments/2013-09/04/c_117215681.htm. Accessed 13 Nov 2014.

deletion and 'transmission-forbidden', on the 9th September 2013 the Judicial Interpretation issued by the Supreme People's Court and the Supreme People's Procuratorate stipulated that if one piece of "libel information" is browsed more than 1000 times or transmitted 500 times, then the poster or re-posters shall be convicted as guilty. As a result, Sina Weibo has quietened down a lot politically, but still boomed economically, while the Sohu Weibo service was "voluntarily" closed down in the late 2014. Tencent's WeChat,[28] an instant messaging program used by hundreds of millions, to which many Weibo users had migrated, appeared to be a flash in the pan, because the authorities have also given it their attention. In May 2013, Tencent intensified efforts to verify the identities of users behind public accounts. And in August, these restrictions were formalized when the government prohibited instant messaging accounts from posting political news without official approval. As the latest development, Chinese government has started to advertise its conception of Internet sovereignty to international society, arguing that national governments should have the right to supervise, regulate, and censor all electronic content transmitted within their borders. The conception emerged as a foundational policy following the establishment of a high-level organ on 27 February 2014, Central Leading Group for Cyberspace Security and Informationization Affairs. This emerged in the wake of Edward Snowden's revelations that the U.S. had been snooping on the Chinese Internet for years.[29]

Being aware of the advantages of Weibo, the government also managed to put them to its own advantage by setting up governmental Weibo accounts. By the end of 2012, among the total 309 million of Chinese microblog accounts, over 176,000 were opened and operated by Chinese government agencies, which have been used to disclose government information and mediate relationships between the governed and those governing (Zheng 2013; Criado et al. 2013). Moreover, the State Council issued the Guidance on further Improving the Openness of Government Information to Respond to Social Concern and Enhance the Credibility of the Government on October 15, 2013, requiring that all levels of government departments "actively explore the utility of such new media as Weibo and WeChat, release authoritative government information on a timely basis, especially that involving major public concern over public events as well as policies and regulations, and make full use of new media's interactive features to interact with the public on a timely basis".[30]

---

[28]Launched in early 2011, WeChat combined the function of social networks and instant messaging, closely matching netizens' need for quick interaction. Designed specifically for mobile devices, it had reached 50 million users by the end of the year. It broke 200 million in 2012, and cracked 300 million in early 2013. Functionally different from Weibo, WeChat is geared more towards one-on-one chat than broadcasting your thoughts to an army of followers.

[29]Livingston, S. D., Beijing Touts 'Cyber-Sovereignty' in Internet Governance, http://www.chinafile.com/reporting-opinion/viewpoint/beijing-touts-cyber-sovereignty-internet-governance. Accessed 19 Feb 2015.

[30]http://politics.people.com.cn/n/2013/1015/c1001-23204203.html, October 15, 2013. Accessed 15 Jan 2015.

In short, the emergence of personal media has brought a large degree of civic participation in social-economic-political affairs in China (Mou et al. 2013). Increasing online dissemination of mass incidents and the strength of combined public opinion have highlighted the power of civic forces. Although the government still has a privileged position in governance of the Internet, the space for freedom in Chinese society has to a degree been enlarged.

The Logic of Internet Politics with Chinese Characteristics

Although Internet service has faced constant interference from China's army of online censors, vigorous public debates still occur in the Internet space, and has played an increasingly important role in anti-corruption, legislative reform and the practices of discursive democracy. Underlining such practices, the logic of Internet Politics in China can be discerned.

Dynamics of Political Change in China

Social stability has been the priority of the CPC and the government. By some standards, it is most unlikely that China will transform politically, because it has a so-called centralized party-state system: the power of the communist party and the government is ubiquitous, controlling the ideology, the army, the judicial system, almost everything. How, in such a system, can you imagine that any political change might emerge? Where is the source of such a change?

We argue that it is in such a system, through the implementation of the Internet, that strong dynamics encouraging change will arise. Generally speaking, diversity is the source of social transformation. It is well known that China has a strictly hierarchical governance structure along with a hierarchical information structure, where the central committee of the CPC is at the center of power. However, China's political structure is fragmented, as has been noted by a number of economists and sociologists (e.g., Xu 2007; Zheng 2008). At times, the political games between the party and government and between central and local government are so tricky that they may become catalysts for political change, and even for economic development (Cheung 2009). For example, conflicts of interest between different governance bodies (and the related operational entities) have meant that tri-networks integration, the convergence of telecommunication, Internet and broadcasting networks, has taken so many years and has still not been fully achieved. As a result, the voices for reforming the related systems have been growing in strength.

In fact, Chinese leaders and officials at all levels of government always live in an ambivalent situation. As far as the Internet is concerned, on the one hand, they need to promote the development of information technology, for "science and technology is the first productive force"; on the other hand, they know clearly that encouraging the development of the Internet is almost certain to produce unintended consequences such as social protests and an upsurge in collective actions. China has therefore established a parallel governance structure: one is used to promote the development of the Internet industry (mainly the Ministry of Industry and Information Technology, the State Internet Information Office, and there are corresponding authorities or departments at the provincial or lower levels), and the other for the implementation of political control (mainly the Central Propaganda Department, and there are also corresponding departments at the provincial or lower levels). The management of the media is actually executed by the Communist party's propaganda department, on the basis that the media shall be consistent ideologically with the party, and shall propagandize and abide by the guidelines and policies of the party. Basically, the two mechanisms are run by two groups of people, and the State Administration of Press, Publication, Radio, Film and Television (there are also corresponding authorities or departments at the provincial or lower levels), directly under the State Council, is by and large positioned at the crossover point. Execution of such governance on the Internet is not always consistent, and there is undoubtedly an inherent tension (see Zheng 2008).

Moreover, the government at times needs to mobilize social forces through the Internet to support their policies. On the one hand, governments, especially the central government, have an incentive to encourage exposure of specific forms of accidents or illegal behavior, such as mining accidents or corruption, related to lower level authorities, to enhance the legitimacy of the ruling elites by coping with them strategically. On the other hand, when differences in opinion over certain issues emerge within the central leadership, both the reformers and the conservatives have an incentive to take advantage of the Internet to mobilize support from the public. For example, although there were controversies over the Regulation for Custody and Deportation of Urban Vagrants among officials long before Sun Zhigang's death, it was after the incident occurred and had been fiercely discussed on the Internet that the reformers were given the opportunity to put it on the agenda and eventually to abolish it. Similarly, as early as 2005, the aim of abolishing the laojiao system[31] was listed in NPC's annual legislative plan, which was put aside for 8 years because of lots of disagreements. Only with the public support aroused by

---

[31] Adopted in 1957, the laojiao system, or re-education through labor, is a system of administrative detentions which is used to detain persons for minor crimes such as petty theft, prostitution, and trafficking illegal drugs, as well as religious or political dissidents such as FalunGong adherents. Laojiao sentences are typically for 1–3 years, with the possibility of an additional 1-year extension. They are issued as a form of administrative punishment by police, rather than through the judicial system. See, http://en.wikipedia.org/wiki/Re-education_through_labor. Accessed 15 Dec 2014.

the Tang Hui[32] and Ren Jianyu Incidents[33] did Chinese leaders make up their mind to abolish it.

Furthermore, the contradiction between rhetoric and reality is also a major source of political change in China. There are remarkably pleasing slogans like "socialism is to liberate the productive forces", "to pursue common prosperity", "to serve the people wholeheartedly", and "the Party is built for the public and it exercises state power for the people", and so on. In this way, any gap between rhetoric and reality may motivate change in either of them. Be they on left or right, protesters and reformers can always be motivated to change their politics in the face of such contradictions. With the rapid economic development since the implementation of the reform and opening up policy in 1978, Chinese people's living standards have improved tremendously. However, a variety of social problems such as increasing inequality, abuse of power and ubiquitous corruption, have given rise to a great deal of resentment and hatred, which has resulted in lots of protests and collective actions. In this sense, the dynamics of political change has always been there. Actually, both the rise and downfall of Bo Xilai stemmed partly from his mobilization of the new left, made up of both Maoists and social democrats disillusioned with the country's market-based economic reforms and increasing economic inequality.[34]

What Difference Does the Internet Make in Chinese Politics?

In China, most social resources, including information resources, have been controlled by the party-government. With the development of the Internet, control over the dissemination of information on public issues has been loosened, although the government still regulates news reports and information discourse online.

Within the hierarchical governance structure, for officials, it seems to be normal to conceal the true state of affairs from above and below themselves, and to report only what is good while concealing what is unpleasant. In the meantime, the public can only access the processed information, either from the central or local media, the mouthpieces of the party and governments. Moreover, in line with the propaganda discipline, local media have no rights to capture and report news from other

---

[32]Tang Hui, mother of a kidnapping and rape victim in Yongzhou, Hunan province, received a laojiao sentence for repeatedly petitioning for harsher sentences for her daughter's attackers. See, "Tang Hui", http://baike.baidu.com/. Accessed 15 Dec 2014.

[33]On August 18, 2011, Ren Jianyu, a 25-year old college-graduate village official who was about to become a formal civil servant, received a laojiao sentence with the charge of "Incitement to Subvert the State Power". The evidence was those hundred plus "negative comments and information" about incidents of pollution, corruption in public authority, and domestic political reform that he had posted or transmitted in his QQ space, Weibo and some public BBS. See "Ren Jianyu Laojiao Case", http://baike.baidu.com/. Accessed 17 Dec 2014.

[34]China's new leaders: The princelings are coming. *The Economist*, June 23, 2011.

regions. In this way, a lack of transparent information all over the country seems to be the norm, and the abuse of power and corruption is destined to be inevitable. For example, at the beginning of outbreak of SARS in China, the relevant information was masked for various reasons, and the public had to obtain information through SMS, email or BBS. Lack of authoritative information led to the breeding and spreading of rumors, and the government's credibility was greatly affected. With the wide application of the Internet, especially of Web2.0 technologies, however, a universal information sharing platform has been established, which means the authorities are unable to fully control the disclosure of information, and anyone at any time can receive and publish information on the Internet regardless of her geographical location. Especially with the personal media like Weibo, anyone has the opportunity to disseminate information. In this way, the Internet has the potential to disclose sensitive information concealed by the authorities, especially local authorities, and rapidly to transform a local event into a global one, so as to play watchdog. The Internet, especially its updated application, Weibo, has become the best choice, because everybody can be a medium, even stronger than a popular newspaper or TV station. As a result, the public voices, laden with much political pressure, can easily draw attention of decision makers at the upper levels, which may eventually result in political effects. In enabling netizens' voices to be heard, the Internet has propelled the government towards transparency and accountability, and has even helped to improve laws and regulations.

In the meantime, with new mechanisms for the dissemination of information, the gap between rhetoric and reality has been increasingly highlighted in the mind of the ordinary people, and this has produced even greater social discontent. In fact, the image of China in government-owned media and that on the Internet is always in sharp contrast. The government-led media has been organized to sing the praises of the party and government, showing that this country is in the heyday of peace and prosperity, while the dark side of the society is exposed unavoidably through the Internet due to its diversity and competitiveness in news reporting, which is more likely perceived by netizens as much closer to the truth. In this way, with the popularity of the Internet, netizens even no longer believe what the official media report and what the authorities say. Thus, people would have stronger reasons to seek changes in their lives and to participate in the process of collective political action. In addition, being exposed to various kinds of information, common citizens have profoundly altered their perspective on public affairs. This has further promoted the reforming process in China (Li et al. 2013).

Moreover, with the Internet, the persecuted, protesters and reformers have more opportunities to access resources and receive moral support (Li and Liu 2014), and the transaction costs involved in collective actions mediated by the Internet have declined dramatically. It is much more convenient for ordinary people to take or to be involved in collective actions with common political appeals. Although the freedom of association is people's fundamental right according to the Constitution of China, it is often difficult for it to be implemented in reality. The Internet has technically empowered people to associate themselves for their common goals. This seems to have produced a serious situation for the government. Furthermore, the Internet as a platform has fostered a special kind of collective action, online crowdsourcing,[35] to be used to deal with socio-political affairs. Weibo-based online crowdsourcing usually consists of four phases. The first is the initiative phase. The initiator can be individuals, media, even government agencies, whose purpose is to make their voices heard. With mobile Internet, any user can formulate a topic, and any others are free to express their opinions about it, and the topic will spread far and wide quickly, which leads to the second phase: consensus building. In such a process, each actor behaves the same, posting and transmitting, with the purpose of emotional mobilization and target-searching, and the e-influentials will emerge at this moment, mostly celebrities or organizations like the media. The third phase is the formation of the solution along with the consensus reshaped under division of labor among the netizens. In the final phase the collective action is closed if the issue is handled properly, for example those responsible are punished, victims receive compensation or other appropriate arrangements are made. A good illustration of this is the case of Yang Dacai[36]: through online crowdsourcing, an official who showed up at a crash scene was translated firstly as the "smile brother", then "watch brother", and finally a greedy official.

The Internet is thus not only the place where people can express themselves, but also the place where they can organize together for common political appeals. In this way, the Internet is destined to exacerbate social instability by intensifying the netizen's perception of the gap between rhetoric and reality, and by endowing them

---

[35] The concept of crowdsourcing was initially put forward in the business world. It is defined as the act of outsourcing tasks originally performed inside an organization to a large heterogeneous mass of potential actors (e.g. Howe 2008).

[36] At 2:18 AM on Aug. 26, 2012, a long-distance bus carrying 37 passengers collided with a tanker loaded with highly flammable methanol on a Chinese highway in Shaanxi Province. Both vehicles burst into flames, killing 36 passengers. Pictures of the accident began to circulate on Sina Weibo. Soon the attention of Weibo users was drawn not to images of the accident itself but to a person who had shown up at the crash scene, grinning amid the wreckage shortly after the tragedy. He is Yang Dacai, chief of the Shanxi Provincial Work Safety Administration. As Weibo users tried to figure out whom this strangely smiley official was, pictures of Yang wearing luxury watches spread like wildfire. Weibo users rallied to collaborate online collaboration to uncover his information. Yang was quickly dubbed "the smiling brother," and his laughing visage became September's most comical Internet meme. One day after the accident, Weibo users posted five photos of Yang wearing five different luxury watches, including a $63,000 Vacheron Constantin and a $10,000 Rolex. Many netizens questioned how a government worker who would have not been making more than $15,000 a year could afford so many expensive watches on his salary. In a nation struggling with rampant corruption, Yang soon acquired another nickname: "watch brother." On Aug. 29, Yang apologized for his unfortunate behavior at the scene of the accident. "In fact I was not smiling at that time," he wrote on his Weibo account, "My facial expression was a little relaxed. What I did was just to make my colleagues feel more comfortable." Yang reiterated that he bought these five luxury watches with savings accumulated over a 10-year period. However, Yang's explanation did not dispel suspicion. On August 30, Sina Weibo detectives found photos of Yang wearing another four luxury watches. On the same day, the Shaanxi Provincial Discipline Committee told state news media that it would launch an investigation of Yang and publicize the result of the inquiry. On Sept. 21, Yang was relieved of his position and accused of serious discipline violations.

with resources and moral support for collective actions. The Chinese authorities are clearly aware of the potential danger of such organized actions in terms of social stability. In fact, a major reason for the government crackdown on the Falungong is for its strictly-organized features mediated by modern information technology. It was often said that the old China, before the year 1949, was in a state of disunity like loose sand and thus was not strong enough to resist foreign aggressions. The Communist party clearly understood the chronic sickness of Chinese society and has devoted itself to uniting people around its banner, and at the same time never allows spontaneous associations with political objectives. In fact, online posts containing only negative, even vitriolic, criticism of the state, its leaders, and its policies are not more likely to be censored, and censorship on the Internet is aimed at curtailing "collective action" by silencing comments that represent, reinforce, or spur social mobilization, regardless of content (King et al. 2013). This is why the government wanted to crackdown on opinion leaders involved in the organized, for-profit or ideal-motivated political actions mediated by Weibo in mid-2013. Of course, not every collective action mediated by the Internet would be curtailed. Whether certain collective actions succeed or not depends largely on the strategy the netizen takes (Zheng 2008). Actually, compatibility of incentives is essential for the success of collective action online, which means that what the actors are asking for has to be legitimate and in accordance with the interest of the leadership at the highest levels. At least, it should not be perceived by the whole leadership as against them or the current regime. Reformers in the leadership might have reasons to deal with the issue cooperatively. Fortunately, in the current legal framework, one can always find certain legal provisions to support one's appeals. So long as the protestors act in accordance with them, the odds on success will be high.

What Difference Do Chinese Politics Make to the Internet?

As a set of services, platforms, standards, and user behaviors, the Internet is also shaped by political conditions. As Morozov (2014) argues, it has never been and never will be the "same Internet", even in the context of a single country. In other words, the Internet is still in its reifying process, generating a variety of possibilities (Feenberg 2014). Indeed, the Internet in China is very different to that in the West. Such differences partly lie in the political structures underlining the development of the Internet. In this sense, it is more appropriate to discuss "internets" rather than the Internet.

To support such an assertion, let us look at again the development of Weibo services in China. Weibo, as "the most powerful media" in China,[37] has helped to

---

[37]Li, Kaifu, Weibo: the most powerful media in China. http://www.linkedin.com/today/post/article/20121025152902-416648-weibo-the-most-powerful-media-in-china?goback=%2Egde_1398377_member_186558682. Accessed 10 June 2014.

constitute a relatively free speech environment and even a public sphere for the Chinese, although it is strictly governed by various censorship policies. Actually, Weibo is functionally very different from Twitter and Facebook, which accounts for its popularity in China. According to a report by Chinese Academy of Social Sciences (CASS), over 70 % of Weibo users take Weibo as their primary source of news, and some 60 % believe it is trustworthy. Nearly 60 % respond they are more willing to express their political views on Weibo. In contrast, only about 9 % of Americans respond that they get news mainly from microblogs, according to a Pew Center report.[38] Another survey conducted jointly by UCLA and CASS, showing that 61 % of Internet users in China believe that the Internet empowers them to comment on the government's behavior, contrasts with only 20 % in the US, 24 % in Japan, and 26 % in South Korea (quoted by Zheng 2008: 118). Such a contrast is mainly due to the different political structure of China and the West. In the West, if people encounter problems, there are legislative systems, elected representatives and independent courts to seek help, and there are also many channels for them to make their voices heard other than the Internet. In China, however, the legal system is not very sound, and to make things public through the media seems to be the last resort for many discontented Chinese. Moreover, it is widely known that the traditional media is strictly regulated in China, and there are few formal channels for the ordinary people to express their views. Once the Internet is there, the discontented swarm into it and have their voices heard. Just as blind people develop compensatory mechanisms for their loss of vision, the Weibo service has developed a unique configuration that has attracted so many Chinese people to "enroll". Topics like social justice and civil rights, the corruption of officialdom, environmental pollution, and reform of the political system have always been the foci of Weibo users. In fact, considering how large China's population is, such a situation implies enormous business opportunities. With just this foresight, ISPs such as Sina and Tencent have developed their Weibo or WeChat services in their unique styles, and have achieved strong competitiveness in the market. As a result, there has been a remarkable intersection of the business and political potential of Weibo in China.

According to the whitepaper *The Internet in China*, the Chinese government "advocates the rational use of technology to curb dissemination of illegal information online" (Information Office of the State Council of the People's Republic of China, 2010: Part IV). Undoubtedly, as far as the Internet in China is concerned, the Great Firewall is taken as such a "rational use of technology", made necessary by Chinese politics. As the main instrument of Internet censorship, the Great Firewall blocks users from viewing selected websites, and filters keywords out of searches initiated from computers or smart-phones located in China. Using proxy servers outside the firewall, however, determined users can circumvent the censorship. If users have a secure VPN or SSH connection method to a computer outside China, they can even circumvent all of the censorship and monitoring. Once such circumventions occur, many Internet connections will be further subjected to deep-packet

---

[38] Xuyang and Jingjing, Weibo vs. Twitter, *Global Times*, 2011-5-18.

inspection, where data packets are scanned before being allowed to pass. To evade deep-packet inspection, a method is to forward traffic to TCP 443, obfsproxy and other tools, but it also has been "killed". In short, to every action there is an equal and opposite reaction. With the expected tighter regulation on the Internet, it is expected that technologists will continue to develop new tools to meet the demands of users for reaching blocked services. Behind the Great Firewall is the Chinese conception of cyber-sovereignty, which seems to be a challenge to the Western assumptions about a free and democratic Internet. Unsurprisingly, there is a joke that China has only an "Intranet" rather than the Internet.

Moreover, in order to satisfy the government's requirement that "basic telecommunication business operators and Internet information service providers shall establish Internet security management systems and utilize technical measures to prevent the transmission of all types of illegal information", ISPs in China have to conduct their own censorship for the government, employing an army of censors to trawl content to remove offending materials. In addition, the fact one is being censored online leads users to curtail speech and practice self-censorship, weighing up carefully the outcomes of everything they want to express online. This may be more effective in blocking internet content than the Great Firewall itself.

Indeed, in the centralized system, the internet seems to have been captured by the central power. With the assistance of the Internet, government has been empowered to enhance itself, and individuals may find themselves under tighter surveillance and control (Feng and Guo 2013). Presumably, with the development of the big data analytic technology, certain individuals' personal data online could be collected extensively, personal footprints could be traced, and even identities reconstructed, which could be used to anticipate and even control their behavior. There would be no personal freedom in such an Internet-based society. In this respect, Minzner (2011) shows great concern about China's future when he argues that the central Chinese authority has chosen to prioritize the party's control at the expense of building autonomous legal and political institutions. Similarly, Sullivan (2014) argued that even though the popularization of Weibo in China has greatly challenged the state's regime of information control, as it enables netizens to publicize and express their opinions online, the government has quickly adapted to the changing network ecology and tried to utilize the Weibo space to its own advantage. Indeed, as the dominant regulator of the Internet, the government appears to have stronger adaptability and resilience than any other actors in relation to constantly changing network technologies. Speech control seems to be more and more effective with the development of the censorship.[39]

In short, Chinese politics has made differences to the Internet in China. On the one hand, China has devoted its efforts to controlling both online and offline

---

[39]Zhu et al. (2013) have researched deletion of posts on microblogs in China, finding that deletions happen most heavily in the first hour after a post has been submitted, that nearly 30 % of the total deletion events occur within 5–30 min, and that nearly 90 % of the deletions happen within the first 24 h.

information which is suspected of being a danger to social stability. On the other hand, online service providers, in fear of being closed down, have to adopt technical means and hire moderators to monitor user-provided content. This kind of censorship and self-censorship has made the Internet in China very different from that in the West. The so-called universal association through the Internet is still undoubtedly a dream.

Conclusion

It has been made clear that, although the Internet has made differences in Chinese politics, it has not brought universal association, and sufficient equality, freedom and democracy automatically, and certainly politics has not disappeared in China. Rather, the Internet has been shaped by Chinese politics and thus given rise to quite different ISPs and netizens in China. In this sense, the paradise image of the Internet does not hold, although it could still be accepted that the Internet's great potential has been subverted by "political centralization".

Alternatively, the situation in China seems to be a panopticon, to borrow Michel Foucault's analysis. To him, as a highly efficient surveillance mechanism, the panopticon can be discerned in institutions in modern society such as psychiatric clinics, boarding schools, and military camps. With the development of the Internet, the Chinese authorities have been empowered by the Internet, and censorship on the Internet has advanced to such an extent that individuals have been under tighter surveillance and control. So the image of the Internet as the panopticon appears to be closer to the reality in China (Liu 2014). As a revised version of the panopticon, Yu (2009) came up with the concept of "onlooking prison" to describe the structure of the Internet in China, where regulators lose their full control of information resources, and the public are able to monitor the behavior of the "guards" and to have their appeals heard and satisfied.

However, a more open and neutral image of the Internet is the laboratory, which is in accordance with the experimental metaphysics developed by Bruno Latour. For him, the world is just "an immense, messy, and muddy construction site" (Latour 2004: 161), and there is no a priori determinant of who are the main actors and their properties; due to complicated translations, nothing should be decided or assumed in advance. From such a perspective, society is nothing but a laboratory, a facility that provides conditions in which various experiments and measurements may be performed. In his article on laboratory studies, Latour argued that there are full transformations in the laboratory, and "the categories of inside and outside are totally shaken up and fragmented by laboratory positioning", so it is "in laboratories that most new sources of power are generated", and "the very content of the trials made within the walls of the laboratory can alter the composition of society" (Latour 1983: 153, 160, 159). Similarly, it is in the Internet that most new sources of power are generated, and the very content of experiments by such actors as academics, entrepreneurs, journalist, netizens, politicians etc., even including the nonhuman, within the Internet have altered the composition of society. Therefore, we can also parody Archimedes, "Give me the Internet and I will move society". To take one more step, it should be noted that the Internet as a laboratory is also in an evolving process, being shaped by experiments conducted by a variety of actors.

Indeed, since the Internet was first introduced in academic circles, different actors have been conducting different experiments in their own way, with the aim of meeting their diversified needs. That's why the public have made efforts to strengthen their disclosure rights and political participation through the Internet, resulting in the appearance of e-influentials, online crowdsourcing and firewall bypassing and so on. Moreover, for-profit organizations, especially ISPs, have been trying to lead the trend or public attitudes and interests by innovating Internet applications, while the government has created strict regulations to govern online behaviors such as setting up the firewall, the real-name registration system, and information filters, as well as opening their own Weibo accounts. Their actions seem to be inconsistent, but coexist in the Internet ecology, resulting in a distinctive Internet with Chinese characteristics, the most complicated political landscape in the history of mankind.

From this perspective, we can make comments on the contribution of Zheng (2008), who has reviewed in some detail the changing relationship between the state and society in China under the influence of the Internet. For him, the Internet has empowered both the state and society, and the interaction between the state and society is in a form of recursion, leading to the transformation of each other. Although Zheng's argument is stimulating and correct to a large extent, basically he still regards the Internet as a black box. The reality is that the state, society and the Internet are all in dynamic interaction with each other, and the Internet itself is in a constantly reifying process due to such interactions. Along with the development of Internet in China, all actors have kept changing their behaviors to fit in, proactively or passively.

Therefore, politics still matters. To see the potential of technical networks is valuable, but to see the right way of realizing such potential is even more important. The transformation of society mediated by networks from the closed situation to universal association is not so much an autonomous process as a series of real social and political choices. What is needed is exactly foresight and political choice, by which a variety of stakeholders in society participate together in discussing and designing their common future. Different from forecasting or prediction (Martin 1995), the concept of foresight presumes that the future is uncertain and that there is more than one possible development path, and that which one may be realized largely depends on collective projection in advance. With foresight, people will be empowered to harness technical networks to serve the interests of their communities better.

Although we cannot determine the outcome of our choices, we should still do our best to make such choices better, as a Chinese old saying goes, "The planning lies with man, and the outcome with Heaven". For China, such a hope lies exactly in the white paper *The Internet in China*, issued by the Chinese government in June 2010, which says: "The Internet provides unprecedented convenience and a direct channel for the people to exercise their right to know, to participate, to be heard and to oversee, and is playing an increasingly important role in helping the government get to know the people's wishes, meet their needs and safeguard their interests. The Chinese government is determined to unswervingly safeguard the freedom of speech on the Internet enjoyed by Chinese citizens in accordance with the law".

References

Castells, M. 2001. *The internet galaxy: Reflections on the internet, business, and society*. Oxford: Oxford University Press.
Chen, W. 2010. On the developing process of internet regulation from government in China. *Today's Mass-media* 10: 112–114.
Cheung, S. 2009. *The economic system of China*. Beijing: China Citic Press.
Criado, I.J., R. Sandoval-Almazan, and R.J. Gil-Garcia. 2013. Government innovation through social media. *Government Information Quarterly* 30: 319–326.
Feenberg, A. 2014. Great refusal or long march: How to think about the internet. *Journal of Engineering Studies* 6(2): 146–155.
Feng, G.Ch., and S.Z. Guo. 2013. Tracing the route of China's Internet censorship: An empirical study. *Telematics and Informatics* 30(4): 335–345.
Howe, J. 2008. *Crowdsourcing: Why the power of the crowd is driving the future of business*. New York: Crown Business.
Kim, W.S., and A. Douai. 2012. Google vs. China's "Great Firewall": Ethical implications for free speech and sovereignty. *Technology in Society* 34: 174–181.
King, G., J. Pan, and M.E. Roberts. 2013. How censorship in China allows government criticism but silences collective expression. *American Political Science Review* 107(2): 1–18.
Latour, B. 1983. Give me a laboratory and I will raise the world. In *Science observed*, ed. K. Knorr, and M. Mulkay, 141–170. London: Sage.
———. 2004. *Politics of nature: How to bring the sciences into democracy*. Boston: Harvard University Press.
Lessig, L. 1999. *Code and other laws of cyberspace*. New York: Basic Books.
Li Q., and Q. Liu (ed.). 2014. *The internet and transitional China*. Beijing: Social Sciences Academic Press.
Li, Q., Q. Liu, and Y. Chen. 2013. The influence of internet on society and its construction. *Social Science of Beijing* 1: 4–7.
Liu, Y. 2014. *Minerva in action: The power dimension of contemporary cognitive activity*. Chengdu: Southwest Jiaotong University Press.
Liu, J., and W. Jin. 1998. *Awakening after thousand years: Information and knowledge economy*. Beijing: Social Sciences Literature Press.
Martin, B.R. 1995. Foresight in science and technology. *Technology Analysis & Strategic Management* 7(2): 139–168.
Minzner, C. 2011. *Countries at the crossroads: A survey of democratic governance (2011: China)*. Washington, DC: Freedom House.
Morozov, E. 2014. *The internet, politics, and the politics of internet debate*. https://www.bbvaopenmind.com/en/article/the-internet-politics-and-the-politics-of-the-internet-debate/?utm_source=views&utm_medium=article10&utm_content=morozov-article. Accessed 3 Dec 2014.
Mou, Y., D. Atkin, H.L. Fu, A.C. Lin, and Y.T. Lau. 2013. The influence of online forum and SNS use on online political discussion in China: Assessing "spirals of trust". *Telematics and Informatics* 30(4): 359–369.
Sullivan, J. 2014. China's Weibo: Is faster different? *New Media & Society* 16(1): 24–37.

Wang, S.S., and J.H. Hong. 2010. Discourse behind the forbidden realm: Internet surveillance and its implications on China's blogosphere. *Telematics and Informatics* 27: 67–78.

Wang, L.Y., W.N. Qu, and X.H. Sun. 2013. An analysis of microblogging behavior on Sina Weibo: Personality, network size and demographics. In *Cross-cultural design: Methods, practice, and case studies, CCD/HCII 2013, Part I, LNCS 8023*, ed. P.L.P. Rau, 486–492. Heidelberg: Springer-Verlag.

Winner, L. 2014. A future for philosophy of technology – Yes, but on which planet? *Journal of Engineering Studies* 6(2): 141–145.

Xiong, X.B., G. Zhou, Y.Z. Huang, H.Y. Chen, and K. Xu. 2013. Dynamic evolution of collective emotions in social networks: A case study of Sina Weibo. *Science China-Information Science* 56: 1–18.

Xu, K. 2007. *Executive force of the government*. Beijing: Xinhua Press.

Yang, Q.H., and Y. Liu. 2014. What's on the other side of the Great Firewall? Chinese web users' motivations for bypassing the internet censorship. *Computers in Human Behavior* 37: 249–257.

Yu, G.M. 2009. Media revolution: From Panopticon to onlooking prison. *People's Tribune* 16: 23–25.

Zheng, Y. 2008. *Technological empowerment: The internet, state, and society in China*. Stanford: Stanford University Press.

Zheng, L. 2013. Social media in Chinese government: Drivers, challenges and capabilities. *Government Information Quarterly* 30: 369–376.

Zhu, T., D. Phipps, A. Pridgen, J.R. Crandall, and D.S. Wallach. 2013. *The velocity of censorship: High-fidelity detection of microblog post deletions*. Paper for the 22nd USENIX Security Symposium in Washington, DC., August 2013. http://arxiv.org/ftp/arxiv/papers/1303/1303.0597.pdf. Accessed 5 Dec 2014.

Zuboff, S. 1988. *In the age of the smart machine: The future of work and power*. New York: Basic Books.

Chapter 8
The Rise of Pirates: Political Identities and Technological Subjectivities in a Network Society

Rodrigo Saturnino

Introduction

Information is power. This is a well-known expression which opens the *Guerilla Open Access Manifesto* (Swartz 2008) – a revised version of the *Declaration of the Independence of Cyberspace*, by Perry Barlow (1996) – written in 2008 by North-American hacker Aaron Swartz. Aaron's life history became known worldwide as a result of his suicide on January 11, 2013, following his 35-year sentence to federal imprisonment for having bulk-downloaded, through the servers of the *Massachusetts Institute of Technology* (MIT), thousands of academic journal articles deposited at JSTOR *(Journal Storage)*. Swartz, who was 26 at the time, described himself as one of the most prominent persons fighting for the importance of the free sharing of knowledge through digital networks as a fundamental feature of social development.[1] Swartz strongly believed that knowledge was the key element in a better society, a key so fundamental that it had to be shared around the world in opposition to the information blockade imposed by publishing monopolies.

Sharing is a moral imperative, a humanitarian act against the privatization of knowledge. In the words of the late *hacktivist*, fighting for an open Internet demands courage. Overruling the laws which prescribe punishment for those who base their practice on a political consciousness of information requires a mindset

---

[1] JSTOR, a digital library founded in 1995, sells much of the material that keeps stored in its servers. The JSTOR file downloading case began in the autumn of 2010, through a guest-user's account. The unusual data traffic between MIT's and JSTOR's servers drew the attention of both institutions, due to alleged system overload. A first attempt was made to block access, but the account was subsequently restored. Massachusetts State Police and the Federal Bureau of Investigation (FBI) were then called in to investigate the situation, and in January 2011 Aaron Swartz was arrested at the MIT's *campus* (Sims 2011; Macfarquhar 2013).

R. Saturnino
Instituto de Ciências Sociais, Universidade de Lisboa, Lisbon, Portugal
e-mail: rodrigo.saturnino@gmail.com

that understands it not merely as a contemporary element in power frameworks, but as a public asset; a right to which every person should be freely entitled.

Internet is politics. Although Aaron's personal tragedy was regarded as the consequence of his long-term depression, there was an immediate response to his death. Protests and demonstrations spread worldwide, not only because he was young, but also because his activism represented an international cause: Internet freedom. In addition to this particular programmer's case, there are several others which reinforce contemporary struggles and involve new protagonists, drawn from the tensions caused by legal attempts to criminalize and intimidate alternative forms of content consumption and distribution (e.g. movies, music, academic articles, videos, images, etc.), generally protected by copyright and intellectual property laws.[2] These activisms, which emerged under the sometimes utopian umbrella of the Internet as being a space for the free sharing of knowledge, remind us how the ambivalence of information, the elasticity of the network idea and the ambiguities of piracy gave rise to a variety of social conflicts at the political, legal and economic levels, involving life's deepest layers.

At the institutional level, the rise of the Pirate Party in many countries is one of the clearest examples of such current conflicts.[3] In a dialogue with the thought of Musso (1999, 2003, 2004, Chap. 2) and the reflections on the social implications of networks as put forward by authors like Mattelart (1995, 1996), Garcia (2006, 2010), and Martins and Garcia (2013), this paper seeks to classify this international movement as an effect ideologically influenced, either by the social diffusion of communication technologies, or by the conversion of information and digital networks into important elements in the struggle for power.

Methodologically, this text makes use of the Pirate Party's trajectory to describe how international conflicts in the field of Internet regulation expanded the cleavages in the network as a transformative device, making it not just a catalytic space for the capitalist dynamic (Schiller 2000, 2014), but also a vortex of new political identities and new global forms of agency (Hands 2011; Postigo 2012). Still in the analytical framework, it discusses how the reticular imaginary helped to redefine the rules of the world economy and forms of consumption, and in the same way, served to mobilize technological subjectivities committed to reorganizing network geopolitics.

---

[2]Noteworthy are not only the examples of Edward Snowden's and Julian Assange's 'Wikileaks' revelations, but also the actions of *Anonymous* and the recent activism of groups such as *Open Access Movement*, *Free Software*, and others.

[3]In the text, the label "Pirate Party" was used to identify all political parties (official or non-formalized) in different countries which share common causes, such as freedom of information and knowledge sharing, the reformulation of copyright and patent laws, the right to privacy, public transparency and direct democracy, amongst others.

Polysemy, Physiognomy and Metaphors: Network Ambivalences

Since the seventeenth century, the idea of *network*, conceived in an analogous manner by philosophical thought, has designated different representative forms in the constitution of economic and social relations (Merklé 2004). The term is a transdisciplinary apparatus, used to qualify both a type of pathway and the gatherings of individuals (Wellman and Berkowitz 1991). One speaks of railway networks, fishing nets, research networks, social networks, computer networks, networks of neurons, etc.

Despite the different applications and the variety of fields that employ it, the word "network" maintains a common core, characterized by complex interconnections of systems that are both physical and imaginary (Sfez 2002). In this sense, the network can be interpreted as a device used to define and organize different ways of thinking about the relationships that are created in the social fabric by means of *decentralized communication* and the *proximity* that is established between the subjects and the elements within it.

The contributions made by Musso (Chap. 2) reiterate the epistemic character acquired by the network, from its beginning as a technical artifact to its convergence towards a social symbol. This transformation, started in the late nineteenth century by the work of engineers and industrialists, was legitimized by the socialization of artificial networks through the projection of organic and reticular images borrowed from the human body. With the proliferation of the first technical networks (electricity wiring, railways, etc.), the modern narrative of the network's transformative role found a comfortable place in the social imaginary, to the extent that it began to be reviewed and reactivated as a major tool for world development on a transnational scale. Musso (Chap. 2) believes that movements for the resurrection of the network's architecture symbolicity, arising throughout history, substantiate their canonical value due to the mythic narratives which were repetitively formed with a view to providing the political and ideological resources which would drive organic transformations in politics, economy and life as a whole.

In the interpretation of Musso (2004), the physiognomy of the network is designed throughout its history so that it establishes parameterizations from the operation and the organization of things, either belonging to the world of nature or the social world. Similarly, its use holds it in the human imagination as a producer of bonds, and therefore, of a specific moral policy, based on its symbolic ability to maintain equilibrium in environments of intense complexity (body, society, computers, the human brain). In accordance with the thought of Mattelart (1995), Musso (2004) justifies the advent of the network as an ideal metaphor for contemporary organization with the incidences that are later found in Saint-Simonian assumptions and Michel Chevalier's thesis. They would thus fit in with, for instance, the scientific projects explaining the functioning of the world which are found in the works of Norbert Wiener, Claude Shannon and Warren Weaver, as deriving from the formalization of the network as mathematical evidence, and therefore as a disposable resource. According to the author, the rational model of communication networks constituted the technical basis for shaping Saint-Simon's thesis, which defined circulation as the condition of life, as a model of good administration and as a *sine qua non* for social change (Musso 2004: 25; Chap. 2).

Chevalier believed that improvements in communication would bring untold benefits for real, positive and practical freedom. In writing the *Lettres sur l'Amérique du Nord*, in which he reported on part of his mission throughout North America, Mexico and Cuba between 1833 and 1835, Chevalier revealed his amazement at the technological advances achieved by the US in the creation of its road network and machinery. In the same way, he attributed to the idea of communication the idealized objective of not only transforming industry, but also and beyond that, bringing social progress. Putting goods and people into circulation, shortening space, distances and time symbolized not only a breakthrough in the economy between different cultures, but a new opportunity to put into practice the ideals of equality and democracy (Mattelart 1996). Much more than just reducing geographies, it was also important to narrow the social gap between one class and another. Unlike Saint-Simon, Chevalier transformed the network into a symbol-object, identifying its development as an *emerging political revolution*. In this sense, his work stands as an important element in the foundation of communication ideology (Musso, Chap. 2). The technical nature of the network would allow not only undifferentiated communication, but also the common democratization of things by means of an *egalitarian circulation*.

In the words of Musso (2004), the unfolding of Saint-Simon's and Chevalier's thoughts led into contemporary theories of communication utopias (Breton 1992) to the extent they set new symbolic outlines for the political role of the network, outlining a new technological vision strongly guided by the desire for balance, innovation and progress. However, as noted by the author, the explosion and the polysemy of the network absorbed less of the metaphorical character and more of the technical format, changing it into a sort of prosthesis lending support to the search and the will for structural changes in society. In this sense the new communication technologies – with the Internet their most important representative –, would, by reactivating the founding myths of the network, consummate the collection of ideas found in the promises of Chevalier (Musso 2004: 35), both from a technical point of view and at the philosophical level. In the physiognomy of communications, the network is an omnipresent apparatus (Musso 2003; Mattelart 1995).

The French philosopher also stresses that, by virtue of the dual nature of symbols, such revivals bring with them representative ambivalences that are not to be belittled (Musso 1999). According to Musso (2004), overvaluation of the network and information as living artifacts in the communication and circulation process of (biological and social) living organisms produces ambivalences when seen from the policy angle. The anamorphosis of the network as an artifact not only changed the social relationship with space and time, but also fulfilled the aim of using technical arrays to objectify a new condition of life, marked by unequal benefits. It allows circulation, which makes it a living, organic and a *sanguineous system*. But it also favors control, surveillance and punishment.

On the one hand, it is celebrated for enabling the free circulation of things, people and information, favoring an environment of democratic progress and greater public transparency; on the other hand, its technical matrix is criticized for providing optimal mechanisms for control, the automation of gestures, for the marketing of affection and therefore for general surveillance. These dualities enrich and challenge theories on the political role of the Internet. For the skeptics, the network centralizes power, imprisons, curbs and punishes, giving rise to reticular despotism. For optimists and supporters of the network, it is a great tool for peripheral work, for decentralized action and subversion. It is network anarchy. However, the author also highlights that, along with the ideological struggles which take place in the social setting of the network, its intense appreciation as a symbol of progress might encourage a business-like logic. It is a new source of profit when understood as the representation of a new market, which is globalized, personalized and accessible from home (Musso 2004: 35). That is network capitalism.

Social Emancipation and Economic Progress: Promises of the Network

Interpretations of the epistemological elasticity of the Internet using the metaphor of the network have produced a series of *essayistic scenarios* in relation to its social role. In modern-day politics, conceptualizations of *The Network Society* are a key element in interpreting the type of society we live in. At its heart, the Internet emerges as the technical foundation for new forms of organization which developed from the 1960s onwards. Writers like Castells (2005) see it as the *spinal cord* of modern-day societies, in that it enables structural forms which overcome the limitations of earlier attempts to promote social organization.

While *The Network Society* is discussed from a variety of viewpoints – some of them technophobic, in that they recommend a technology which is more moral, and less of a threat to Nature, others overly technophile in their rhetoric, in that they celebrate the advent of the Internet as the final destination of a journey back to an *original community* – we must also recognize that, in addition to its impact on profitability, network technology has brought about structural changes and impacted on the sociability of its users, and consequently on their subjectivities (Floridi and Sanders 2005). The Internet is valued, on the one hand, for the positive dynamics of these new subjectivities, and is therefore self-recommending as an essential element in new processes of empowerment and political and cognitive agency. On the other hand, there are analytically rational approaches to the Internet which see technological innovation as an exponential factor in the discourse on progress and the transformation of the global economy. This strengthens the Internet's function as a mature space for new trends in casual labor (Braga and

Antunes 2009; Scholz 2013) and for new forms of transnational consumerism based on the immaterial, symbolic, aesthetic and social nature of knowledge (Gorz 2003; Rifkin 2000). The network (re)asserts monopolies.

Some critical studies in the sociology of communication have used this apparently contradictory dichotomization of the Internet to argue from a traditional point of view which appears to reject the capitalist system and its consequences, given that the Internet has asserted itself as a space which encourages new ways of producing value based on the commodification of information (Fuchs 2014). Such critiques often avoid interpreting network technologies solely through a defeatist point of view based on a kind of techno-panic generated by asymmetrical systems of oppression, in which the subject is forgotten in heuristic terms. However, the tensions produced by the power struggles over their intensive use, and the divisions which arise over their transformative role, make them catalysts for political action. In this approach, the network is a fundamental element in the acceleration of the world economy and in driving the emergence of new trends in labor and consumption. It is also the crucible of new forms of agency and subjectivity, in which the subject/user attempts to leverage his own productive strength on the basis of the way he uses and appropriates it. In Musso's words (Chap. 2), the reification of the network reifies the struggle.

The development of networks and their interpretation as a device for political revolution led to something one could call the *fetishization of information*. In its digitization aspect, it came to be regarded as an elemental force in a reticular society (Braman 1989), serving almost continuously as a providential contribution to sustaining the social value of technology as a key mediator in processes involving the exchange of knowledge. This theoretical and holistic framework was gradually developed alongside the transformations in the symbolic interaction between individuals and technologies, and with the idea of information, it formed the basis for economic policies that helped solidify a *new technological paradigm*. As understood by Castells (2005), such changes are not changes in the structure of human activities, but in the way these activities began to be performed, i.e. using the digitalization of information as a direct source of productive power that characterizes the biological uniqueness of the social ensemble, that is through "our superior ability to process symbols" (Castells 2005: 142).

In the author's assessment, the affirmation of this technological paradigm, dominant in both developed and developing economies, was shaped by five essential characteristics: (a) information is its raw material; (b) social penetration of the New Information and Communication Technologies (NICTs) is inevitable, because information is an integral part of human ecology; (c) NICTs provide the logic of networks, the only material form to harmonize the complexity of relations; (d) NICTs allow a context of flexibility for their recursive ability to rearrange flows; (e) technology provides an environment of strong convergence, guiding its development trajectories as a common goal in the various scientific subjects and in the formation of specific public policies (Castells 2005: 108).

In the midst of the discussion, the idea of *Network Society* established itself as a symbol of an ideal society; a virtually emancipating political construct, recommending the development and uninterrupted use of NICTs as a fundamental means of meeting social needs. Likewise, network symbolism also forged an imaginary around the digital as a cultural code and evaluative ground for ordering the entire social universe, from the world of work to entertainment and the world of emotions. This society, formed by a social duality produced by the network's fracturing effect, as Castells acknowledges, i.e. an effect which polarizes workers and consumers, included others who were excluded by *informationalism* (Fuchs 2007), and becomes part of the social universe as a sustainable means of information-consumption by *mobilizing desire and taste* (Garcia 2010). In this perspective, the transformation of industrial capitalism to its present form is externalized from the realization of a new kind of consumerism.

From Reticular Dream to Technological Nightmare

When applied to the field of economics, the network represents the organizational framework for methods of production. The main change is based on the privatization of Internet access services, starting in the mid-1950s USA, through appropriation by telecom giants (Schiller 2000), and the inclusion of digital information into the classic category of merchandise and its framing as a work object (Bates 1988). The transformation of various symbols of human language into binary codes by Boolean logic provided the technical basis for reducing epistemic obstacles and expanding its commodification capacity. Nevertheless, since information is an immaterial element, how could one assign economic value to a scientific apriorism that has no characteristics of typical merchandise? The direct consequence of its intangible trait would be its reproduction, almost infinitely, at zero cost. This finding did not feature at all as an incentive to future investors, unless copyright and patents laws were applied to it. Even so, as advocated by Arrow (1984), the issue would be doomed to constant constraints and to various risks if one considers two essential aspects: the impossibility of ensuring ownership of empirically intangible assets and the acquisition of a monopoly, which contradicted the very principle of exclusivity. Under these conditions, Arrow (1984) concluded that the only option left for activities related to the production and dissemination of information would be eventually to become public.

The conclusions reached by Arrow (1984) and other authors who reflected on the intangibility of information, such as Bates (1988), Borgmann (1999), Gorz (2003), Schiller (2000, 2007), Dantas (2003) and Garcia (2006), did not prevent the market from turning theoretical impossibilities into technical profits. One of the strategies used by investors to develop the market based on digital information was to establish costs for its distribution process, considering the dependence on a physical medium to be shared. The Internet is still the best example of this process. It created new professions and extinguished old ones. It established new monopolies and strengthened old ones. Most importantly, it allowed consumers to participate in the game of sharing information thanks to its architecture, the technical reproducibility of information and technological development, which enabled domestic access to powerful computers connected through a global network. The reticular dream transformed the network into a technological nightmare through what Boyle (2008) called a *market failure*. If information represents the *blood* circulating within a system represented by the network, preventing its free circulation would then determine the premature collapse of the market, and the end of Saint-Simon's utopia.

Options to ensure the market's survival, which are based on the digital, can be summarized by invoking ownership rights over information, and therefore limiting access to it, appealing to witch-hunting with the enforcement of laws based on tangible property, or even providing ways of reinventing informational capitalism as the creation of new business models that avoid or postpone its complete destruction. Although creative ways of exploiting the digital universe constantly arise, as is the case for the distribution of music, movies and TV series in a direct way (*on-demand*) through paid or free subscriptions, the first option continues to be the standard rule. All kinds of unregulated information-consumption should be punished. All sorts of unauthorized copying should be punished. According to this line of thought, intellectual property applied to the context of the network is constituted as a legal discipline of monopolistic privileges that seems to endanger the public interests defended by people like Aaron Swartz, either through the unquestionable enforcement of law, the development of international treaties (TRIPS, ACTA, SOPA, PIPA),[4] or by surreptitious forms of privacy violation, social exclusion and technical limitation of access (data filters, suspension of *websites* and Digital Rights Management – DRMs). In this sense, information is transfigured into a *political attractor*, as a mediation pole that intensifies the disputes between values and rules, *pari passu*, the market, courts and citizens (Jordan 2015).

Digital Fights and Technological Subjectivities: Network Conflicts

The technical and legal assaults on free sharing of informational goods served in turn as an incentive for the emergence of the first Pirate Party in Sweden (*Piratpartiet*) in 2006. Its recent history was driven by two situations which arose in 2003: the creation of the Swedish Anti-piracy Bureau (*Svenka Antipiratbyrån*), a private agency funded by the Motion Picture Association of America (MPAA) with the objective of safeguarding the application of copyright law in that country, and the emergence of the *Piratbyrån*, a collective initiative by people who wished to put

---

[4]TRIPS – Trade-Related Aspects of Intellectual Property Rights; ACTA – Anti-Counterfeiting Trade Agreement; SOPA – Stop Online Piracy Act; PIPA – PROTECT IP Act (Preventing Real Online Threats to Economic Creativity and Theft of Intellectual Property Act).

the copyright debate on the public agenda, turning it into a political problem (Miegel and Olsson 2008; Sciannamblo 2014).

Taking advantage of the political threats concentrated around the Internet, Rick Falkvinge created, in January 2006, the prototype of what would be the party's website via the Direct Connect hub.[5] In only two days, the *website* had three million visits (Falkvinge 2013: 33). Falkvinge quit his job, took out a bank loan and decided to devote himself full time to creating the party. In the course of its creation, a sudden attack by the Swedish police on the servers of The Pirate Bay (a tracker[6] created by *Piratbyrån* to experience sharing of files indexed in *websites* using P2P technology) resulted in an increase of *Piratpartiet's* popularity. However, the enrollment of thousands of new members did not produce enough votes to win its first election. In the national political election of 2006, the Pirate Party obtained 0.63% of votes. Even without favorable results that would have ensured a seat in the *Riksdag* (the Parliament of Sweden) – with a minimum of 4 % – *Piratpartiet* became the third largest party outside the Parliament, overtaking, for example, the Swedish Green Party (Erlingsson and Persson 2011; Sciannamblo 2014).

When it was created, the main goal of the Pirate Party was to bring copyright to the political debate, questioning its origins and the forms of legitimacy the private sector was granted to monopolize information, consequently restricting civil autonomy in exercising its right to use, for example, peer-to-peer networks. The *Piratpartiet* thus started to stress the revision of crucial concepts for the structural organization and the legitimizing of copyright – such as authorship and intellectual property – in the light of both technological development and the affirmation of moral principles, based on post-materialist values (Miegel and Olsson 2008). Three issues summarized its initial project: reform of copyright law; abolition of the patent system, and respect for the right to privacy.

In the same year the Swedish Pirate Party was founded, activists from Austria, Denmark, Germany, Finland, Ireland, Poland, Spain and the Netherlands founded their own parties, following Falkvinge's initiative. In 2015, the number of countries to raise the pirate flag through their own parties (either formalized and/or in the process of being officialized) surpassed sixty, including countries outside Europe.

The speed with which it spread strengthened the Pirate Party as one of the most significant social phenomena of the twenty-first century, due to something that can be called *ideological franchising*; that is, through *organizational replicas* created in different cultural contexts, as well as the dynamics of internationalization of struggles and dissemination of repertoires globally. As to electoral performance, the Pirate Party is not a success when compared to traditional party machines.

---

[5]Direct Connect is a file-sharing and chat-channels network.

[6]A Tracker (BitTorrent tracker) is a server that assists in the communication between two computers using the peer-to-peer network protocol. A tracker can operate as an Indexer, i.e. the one that also offers a list of files to share. None of them allows direct downloading. The same is true of cyberlockers, which only establish contact between peers.

However, the gradual victories in Europe indicate a progressive penetration in the parliamentary political spectrum, as shown in the Table below:

Global representation of Pirate Parties (2009–2015)

Country	Regional level	Local level	National level	European level	Elected
Germany	45	201	0	01	247
Sweden	0	0	0	02	02
Czech Republic	0	03	01		03
Spain	0	02	0	0	02
Austria	0	02	0	0	02
Croatia	0	02	0	0	02
France	0	02	0	0	02
Iceland	0	01	03	0	04
Netherlands	0	01	0	0	01
Swiss	0	02	0	0	02
Total					267

Source: http://en.wikipedia.org/wiki/List_of_Pirate_Parties. Accessed 5 Feb 2015

In addition to the three basic principles, Pirate Parties' common policies began to favor two other measures: transparency of public administration and the construction of a new democracy. The expansion of programmatic actions received an impulse with their members' reception of the instrumental potential of digital technologies in promoting an effective experience of political action. So they began to defend joint proposals to promote free access to the information on public cases in order to enable not only the inspection, but also the public opening of all cases involving the State. As a final goal, they included a commitment to search for new ways of building a definitely *liquid democracy*, by using the Internet as a popular space for the making of parliamentary decisions.

In the countries which followed Sweden's example, the initial principles have been safeguarded. However, proposals have come to follow a holistically oriented logic, in accordance with different cultural contexts. Ideological unity was reinforced with the creation of the International Pirate Party in 2010 and the European Pirate Party in 2014, establishing an imaginary network, rationally willing to put into practice their associative character in the fight for common goals (Burkart 2014). In this sense, one might think that the internationalization of their actions reflects the rhyzomatic character of the network, and that it represents a framework for the new conflicts that drive the formation of new political identities.[7] Currently, the Pirate Party defines its common policy in a more extensive framework. Although it added basic premises that are also part of the political

---

[7] A comparative table, prepared by Andrew Reitemeyer in an exploratory way, points to the similarities and differences of the policy proposals of the different Pirate Parties. A quick analysis of that document shows that the elementary principles mentioned above remain in evidence in most of the parties listed therein. http://www.cleopolis.com/PP_comparison_policies.html. Accessed 14 Apr 2013.

programmes of traditional parties, themes underlying the technological imaginary (such as openness, freedom, transparency, collaboration, privacy and *sharing whenever is possible*) are still at the center of their ideological case.[8]

From a parliamentary point of view, the Pirate Parties in Sweden, Germany, and recently, Iceland, have achieved a certain formal political respect. However, as marginal parties, they occupy a relatively lower position compared to traditional ones. Although they remain outside the political circles of power, the activism they silently promote seems to expand with actions of social recognition in the production of a new agenda for doing politics in the digital society. The interventions by Swedish Pirate Party's eurodeputies Christian Engström and Amelia Andersdotter, in efforts to reform *copyright* and advocate public openness, for instance, of the closed secret meetings and negotiations around ACTA, which took place in the European Parliament, symbolize these actors' ideological investments in the struggle for a society based on the reticular imagination. Its gradual political acceptance is also signaled, for example, with the appointment of Julia Reda, elected for the German Pirate Party in the European elections of 2014, as the rapporteur in charge of drafting the Parliament's review on the implementation of Directive 2001/29 (known as "Infosoc Directive"), a document devised with the intention of harmonizing certain aspects of copyright law and related rights at the European level. Still in an environment where these actors emerge in the parliamentary scenario, the results of a poll held in Iceland in March 2015 to sound out public opinion on popular support for political parties placed the Pirate Party, led by member of parliament Birgitta Jonsdottir, at the top of the list, with 23.9 % of the population supporting the pirates' causes.[9]

When looking at Pirate Party history and organization, it is clear that the success of networks, namely the Internet, is due not only to its technical ability to maximize the profits of companies which exploit the digital universe and are globally organized through it, but also to the symbolic instrumentality that it provides in the formation of new social dynamics, new forms of association and new forms of

---

[8] In 2015, the common policy of the Pirate parties included the following points: (1) Defend free speech, communication, education; respect the privacy of citizens and civil rights in general; (2) Defend the free flow of ideas, knowledge and culture; (3) Support politically the reform of copyright and patent laws; (4) Have a commitment to work collaboratively, and participate with maximum transparency; (5) Not to accept or espouse discrimination on grounds of race, origin, beliefs or gender; (6) Not to support actions that involve violence; (7) Use free *software*, free hardware, DIY and open protocols whenever possible; (8) Politically defend an open, participative and collaborative construction of any public policy; (9) Direct democracy; (10) Open access; (11) Open data; (12) Solidarity economy, Economy for the Common Good and promote solidarity with other pirates; (13) Share whenever possible. https://en.wikipedia.org/wiki/Pirate_Party. Accessed 10 Jan 2015.

[9] The term "pirate" is used by Pirate Parties to identify their respective members. Despite the semantic ambiguities and the derogatory aspect of the word in contemporary culture, the use made by the Pirate Party members is markedly an affirmative action of the institutional type, in order to mark a place of existence in the field of party politics. http://www.visir.is/the-pirate-party-is-now-measured-as-the-biggest-party-in-iceland/article/2015150318848. Accessed 19 Mar 2015.

political life at local, national and transnational levels. As a network, one can interpret it as a symbol of freedom and economic progress, in line with Musso (Chap. 2), as a means of domination, imprisonment and surveillance, and also as an agency, autonomy and political protest device. The Internet is incorporated as a subsidy for the development of a saturated economy, where information and knowledge form the infrastructure for the readjustment of capitalism, which in turn is founded on the affirmation of economic policies and legal mechanisms for the protection of monopolies. On the other hand, its spread into the spectrum of day-to-day life – exemplified mainly from the 1990s onwards by new social movements and global revolts which took the Internet brand as a catalyst for their actions – reiterates the diversified character given to the social imaginary by the nuances of the network myth (Musso 2004; Postigo 2012).

Today, it is not possible to analyze the Internet while disregarding the collective animosities and technological subjectivities that were formed in everyday life from its intensification as a key feature of the political struggle (Hands 2011). Similarly, it would be naive to ignore the parasitism of companies that benefit from the political and target-oriented ideologies derived from the *emancipatory* function of technology. Current disputes over the domination of the network, represented, for example, by intense party lobbying, behavior manuals and international agreements, and the interests promoting legal frameworks that ensure full control, geopolitical demarcation and censorship, do not remove or invalidate the *praxis* supervening its political use. Nonetheless they do not affirm it either, as a result of the competent prognosis of many authors, who point to the Internet in the expectation of the democratic renewal of a globalized world. Although its action is somehow effective in the field of activism, Internet asymmetries and uses in the world are still guided by the commercial exploitation of its technical quality, which means that users still depend on costly paid access to the infrastructures, which in turn remain in the hands of the elite's telecommunications companies.

To some extent, when thinking about the general relationship between "structure" and "agency" (Archer 1995), these structural features tend to act as elements which diminish individuals' creative capabilities, through a practice that underestimates the interventional variety that may emerge from these agents, both from the middle to the subject as from the subject to himself. Yet the autonomy experiences that arise from it intensify its meaning as a cultural artifact guided by an emotional relationship, which, in turn, feeds its essentiality. Therefore, network polysemy allows one to confront the dichotomous views between objectivism and subjectivism, from a perspective that admits it not simply as a technical object shaped by a functionalist logic – of an action-neutralizing political nature – in which subjects are mere support for monopolies' maximization, but as an instrument for qualitative performances that stand in the collective imagination, less for its *rhyzomatic* technical quality and more for the social and historical character of the creative forms of social interaction which precede its existence.

However tempting this perspective may be, it is implied that it can only be stated as an analytical possibility when interpreted with a bias that distances itself from sociological unilateralism, concerned with framing and structuring the productive historicity of such interactions, in a dualistic and restricted way, and it approaches a *praxeological* analysis, centered in the reasons and in the purposes of conjugations and reflexivities that social agents undertake in order to influence social flows outlining everyday life, whether intentionally or not. In the understanding of Giddens (1979), this ability to transform practice depends on instruments; that is, a "structure" to enable the social agent, and understood not for its historical form to designate the relationships of power, but for its mobilizing function.

In this case, digital piracy in technological networks is paradigmatic. Integrated into the evolving field of Internet regulation, its practice can either be interpreted as a form of harmful appropriation, when it is done with criminal intent, or described as the bias of hedonistic consumerism without any profit, as argued by Lessig (2004). Despite the ambiguities underlying its practice, the main key triggering its political function is the breaking of monopolies. In both cases, although the second is admitted as a new content-distribution mode (Lessig 2004), and therefore something that should be promoted by the law as a healthy refusal of the cultural development of individuals, practice relapses as an act that is legally constituted as a punishable crime, framed under the legal rules governing copyright, intellectual property and patents. However, considering culture transformations as a battlefield and the social conversion of Internet access into a human right[10] together with the varied forms of *electro-consumerism* emerging therefrom, conflicts tend to increase, especially due to the ease of reproduction and circulation of digital information (Boyle 2008).

Pirate Parties worldwide are part of the *evolutionary* and boosting framework that Internet technology provides to those social agents who are in symbolic processes of interaction with the products and goods emerging from the digitization of information. At the same time, it allows adjustment of capitalism through various forms of consumption, both at physical and immaterial levels. The *multitudinous evasiveness* (Negri and Hardt 2001) arising therefrom breaks the issue of intellectual property down into parts, given that the logics underlying policy on the liberalization of flows changes user interaction with the very concept of private property. Everything is accessible on the network. If the network embodies contemporary society, and if information provides the life circulating in it, no one can own the network, because it is impossible for one to own whatever navigates in it. To contain the reticular reflexivities (*sharing is caring*) that mirror users' behavior, the logic of privatizing knowledge, culture and information has operated against policies designed to control and restrict privacy and freedom of Internet use. However, it has also reaffirmed new fields for political fighting (*sharing is fighting*) apparently involved in the conceptualization of the hybrid-mode network; that is, as a public space for political action and a private space for social activity.

---

[10]Through a report by UN's Human Rights Council and the *website* Mashable, the United Nations (UN) defended the access to the computer network as a fundamental human right for the social development of individuals, and recommended the signatory countries to review copyright laws in order to promote a balanced access and without loss to societies. www2.ohchr.org/english/bodies/hrcouncil/docs/17session/A.HRC.17.27_en.pdf. Accessed 10 Jan 2014.

Conclusion

Although the Internet was co-opted as an optimized means for profit maximization, these social movements seem to inaugurate new outlines of social protagonism to be implemented from an embryonic denial of monopolistic and territorial relations of information. The oppositional action that the Pirate Party proposed in the reform of the legal grammar on the laws of intellectual property and copyright reverberates as *practices of resistance* against colonization of the Internet and privatization of information from a discourse that advocates the urgent need to make the network a free and open space (Burkart 2014). The practice they attempt to elaborate is constituted, as demonstrated by Musso (Chap. 2), through the utopian conception of communication networks as tools of democracy, association and equality. Bearing this in mind, its emergence represents an exceptional political moment, protected by the defense of cyberspace as a symbolic place of freedom, transparency, sharing and solidarity. It also reiterates the libertarian and technical-utopian character that once grounded the mythological visions of the network, as highlighted by Musso (1999). The network institutionalizes the fight.

Following the thoughts of Honneth (1996), the action of these actors seems to strive to redress the structural exclusion of fundamental rights (autonomy, privacy, access to culture) from a practice that outgrows the scope of individual intentions, becoming the base for collective and expanded movement. In this perspective, the common struggle for recognition which they seek to establish in the political imaginary, appears to be a claim that goes beyond the party's mere actions to solidify power (McDonald 1999: 155; Honneth 1996). Instead of anchoring in parliamentary procedures making use of a sectored government plan, oriented towards the construction of a private policy, the centrality of their practice in the field of the Internet expands their propostional universe, and depolarizes their field insofar as they use both the homogeneous nature of the law in the process of planning and criminalization of practices that are seen as a threat to economic progress (*we are all pirates*), and the symbolism provided by the network in promoting a transnational community of pairs (*we are all connected*). If the network segregates, surveils and excludes, for the pirates it adds, organizes and empowers.

The external manifestation of these actors' politics seems to be not an act of *resignation*, but rather a *confrontation* with the conflict-resolution aspects and limitations of the law as an ambiguous ground for the management of situations of intolerable illegality, which it differentiates – alongside those which it allows as privileges of the dominant class – in order to formalize, outlaw, isolate or control them. As Honneth (1996) suggests, this movement's commitment reiterates actors' agency, inasmuch as the actor's image as a passive and paralyzed subject is suppressed through a new relationship with himself and through a positive ethics derived from the moral experience of having normative expectations frustrated and digital monopolies disrespected. Conversely, the technological nightmare again becomes part of the pirates' utopian dream. In this respect the political involvement of individuals who are organized by network technological subjectivities marks the institutional rise of a group, perhaps of *a politically specialized minority,* which imagines the future of digital societies emerging from the political renewal of the Internet. That means drawing a new political map containing trails and pathways which strengthen *cyberfreedoms*, ensure digital rights, counteract information capitalism and, above all, create a new identity and a new political category. If the network defines the way the economy is heading and thus where power will reside, for the pirates its rhyzomatic nature favors the creation of a new pathway to digital resilience, political resistance and social life.

This group's trajectory is silent and only peripherally studied. While it is not the final form of political entrepreneurship in the fight for control over the Internet, it signposts a common utopian will to restore collective agency, inasmuch as the network, in its changing forms, as Musso (2004) has pointed out, can be used in a polysemic way, either as an instrument of commercial exploitation or, in an apparent contradiction, as a critical space for the affirmation of identity, for joint recognition, for mutual esteem and, above all, for the public frustration of the artifices which sustain it.

References

Archer, M. 1995. *Culture and agency.* Cambridge: Cambridge University Press.
Arrow, K.J. 1984. *The economics of information.* Cambridge, MA: Harvard University Press.
Barlow, J. P. 1996. *A declaration of the independence of cyberspace* https://projects.eff.org/~barlow/Declaration-Final.html. Accessed 12 Apr 2014.
Bates, B.J. 1988. Information as an economic good: Sources of individual and social value. In *The political economy of information*, ed. V. Mosco, and J. Waski, 76–94. London: The University of Wisconsin Press.
Borgmann, A. 1999. *Holding on to reality: The nature of information at the turn of the millennium.* Chicago/London: The University of Chicago.
Boyle, J. 2008. *The public domain: Enclosing the commons of the mind.* New Haven/London: Yale University Press.
Braga, R., and R. Antunes. 2009. *Infoproletários.* São Paulo: Boitempo Editorial.
Braman, S. 1989. Defining information: An approach for policymakers. *Telecommunications Policy* 13: 233–242.
Breton, P. 1992. *L'utopie de la communication: Le mythe du village planétaire.* Paris: La Découverte.
Burkart, P. 2014. *Pirate politics: The new Information policy contests*, The Information Society Series. Cambridge, MA/London: The MIT Press.
Castells, M. 2005. *A sociedade em rede*, vol 1, 8th ed. Rio de Janeiro: Paz e Terra.
Dantas, M. 2003. Information and labor in contemporary capitalism. *Lua Nova: Revista de Cultura e Política* 60: 5–44.
Erlingsson, G.Ó., and M. Persson. 2011. The Swedish Pirate Party and the 2009 European Parliament election: Protest or issue voting? *Politics* 31(3): 121–128.
Falkvinge, R. 2013. *Swarmwise: The tactical manual to changing the world.* North Charleston: CreateSpace Independent Publishing Platform.

Floridi, L.; Sanders, J.W. 2005. Internet ethics: The constructionist values of homo poieticus. In *The impact of the Internet on our moral lives*, ed. R.J. Cavalier, 195–214. New York: State University of New York.

Fuchs, C. 2007. Informationalism. In Encyclopedia of governance, ed. M. Bevir, 446–448. London: Sage Publications.

———. 2014. *Digital labour and Karl Marx*. New York: Routledge.

Garcia, J.L. 2006. Biotecnologia e biocapitalismo global. *Análise Social* XLI (181): 981–1009.

———. 2010. Tecnologia, mercado e bem-estar humano: para um questionamento do discurso da inovação. *Alicerces: Revista de Investigação, Ciência e Tecnologia e Artes* 3: 19–31.

Giddens, A. 1979. *Central problems in social theory: Action, structure, and contradiction in social analysis*. Berkeley: University of California Press.

Gorz, A. 2003. *L'immatériel: Connaissance, valeur et capital*. Paris: Editions Galilée.

Hands, J. 2011. *@ is for activism: Dissent, resistance and rebellion in a digital culture*. London. New York: Pluto Press.

Honneth, A. 1996. *The struggle for recognition: The moral grammar of social conflicts*. Cambridge, MA: The MIT Press.

Jordan, T. 2015. *Information politics: Liberation and exploitation in the digital society*. London: Pluto Press.

Lessig, L. 2004. *Free culture: The nature and future of creativity*. New York: Penguin Press.

Macfarquhar, L. 2013. Requiem for a dream. *The New Yorker*.http://www.thebuddhasaidiamawake.com/wp-content/uploads/2014/01/Larissa-MacFarquhar-The-Tragedy-of-Aaron-Swartz-The-New-Yorker.pdf. Accessed 20 Mar 2015

Martins, H., and J.L. Garcia. 2013. Web. In *Portugal social de A a Z: Temas em aberto*, ed. J.L. Cardoso, P. Magalhães, and J.M. Pais, 285–293. Paço de Arcos: Impresa Publishing l Expresso.

Mattelart, A. 1995. *Histoire des théories de la communication*. Paris: La Découverte.

———. 1996. *The invention of communication*. Minneapolis: University of Minnesota Press.

McDonald, K. 1999. *Struggles for subjectivity: Identity, action and youth experience*. Cambridge: Cambridge University Press.

Merklé, P. 2004. *Sociologie des réseaux sociaux*. Paris: La Découverte.

Miegel, F., and T. Olsson. 2008. From pirates to politician: The story of the Swedish file sharers who became a political party. In *Democracy, journalism and technology: New developments in an enlarged Europe*, ed. N. Carpentier et al., 203–217. Tartu: Tartu University Press.

Musso, P. 1999. La symbolique du réseau. *Quaderni* 38(1): 69–98.

———. 2003. *Critique des réseaux*. Paris: PUF.

———. 2004. A filosofia da rede. In *Tramas da rede: Novas dimensões filosóficas, estéticas e políticas da cognição*, ed. A. Parente, 17–38. Porto Alegre: Sulina.

Negri, A., and M. Hardt. 2001. *Empire*. Cambridge, MA: Harvard University Press.

Postigo, H. 2012. *The digital rights movement: The role of technology in subverting digital copyright*. Cambridge, MA/London: The MIT Press.

Rifkin, J. 2000. *The age of access: The new culture of hypercapitalism where all of life is a paid-for experience*. New York: Tarcher/Putnam.

Schiller, D. 2000. *Digital capitalism: Networking the global market system*. Cambridge, MA/London: The MIT Press.

———. 2007. *How to think about information*. Illinois: University of Illinois Press.

———. 2014. *Digital depression: Information technology and economic crisis*. Chicago: University of Illinois Press.

Schofield, J. 2013. Aaron Swartz Obituary. *The Guardian*, January 13, sec. Internet. http://www.theguardian.com/technology/2013/jan/13/aaronswartz. Accessed 12 Apr 2014.

Scholz, T. 2013. *Digital labor: The Internet as playground and factory*. New York: Routledge.

Sciannamblo, M. 2014. The internet between politics and the political: The birth of the Pirate Party. In *Piracy: Leakages from modernity*, ed. M. Fredriksson, and J. Arvanitakis, 177–194. Sacramento: Litwin Books.

Sfez, L. 2002. *Technique et Idéologie: Un enjeu de pouvoir*. Paris: Seuil.
Sims, N. 2011. Library licensing and criminal law: The Aaron Swartz case. *College & Research Libraries News* 2(9): 534–537.
Swartz, A. 2008. Guerilla open access manifesto. https://archive.org/stream/GuerillaOpenAccessManifesto/Goamjuly2008_djvu.txt. Accessed 12 Apr 2014.
Wellman, B., and S.D. Berkowitz. 1991. Introduction: Studying social structures. In *Social structure: A network approach*, ed. B. Wellman, and S.D. Berkowitz, 1–14. Cambridge: Cambridge University Press.

Chapter 9
Final Note: Examining the Network Concept

Pierre Musso

As engineers' reticular techno-utopia tinkers with the images produced by Saint-Simonian concepts, the technology of the reticular mind is marking the final stage in the degradation of the network concept developed by Saint Simon. Such is the dual contemporary process of symbolic and theoretical deterioration of the network. The technology of the mind, understood as a canonical process of reasoning used in various disciplines, is an expression of the theoretical dispersion and commercialization of the concept that, with the obligation to think and be networked, has become a "precept". This degraded concept is a catch-all. While it may indeed be useful in various disciplines, that fact of being applicable to everything makes it devoid of all substance. The common denominator is the reduction of the network to the hidden structure of a system, an architecture that can be formalized, made of interlinking relations or inter-connections. This structure is tending to be seen as the universal key to explain the functioning of a complex system of any kind (society, brain, body, planet, world, etc.). Conversely, it is enough to identify or imagine this type of network architecture in, or under, a complex system to infer this system's mode of functioning or transformation. The network defines a hidden order that can be acted upon. This technology of the reticular mind is complementary to the techno-utopia of engineers presented in my introductory text. It reveals the difficulty of conceptualizing the network in any way other than as a metaphor or else reduced to an explanatory structure of a system. The extension of the notion is such that to track whatever remains, I propose that we characterize its constituent parts as fragments of a dilapidated concept.

The sociological approach to networks is marked by the prior description of a society fragmented by the atomization of actors or even individuals (*The Society of*

P. Musso (✉)
Rennes 2 University, Rennes, France

Télécom Paris Tech, Paris, France
e-mail: pierre.musso@telecom-paristech.fr

Individuals by Norbert Elias), immediately compensated for by over-valuing "intermediaries" and by a determinism of relationships and interaction. This "social networks" approach consists of three phases: first, the description of a state of social and institutional fragmentation; then an analysis that values all kinds of ties, intermediaries and relations; and, finally, a formalization of these relations providing a key to interpret society, through the intensity of relations observed between the atomized actors. As Pierre Bourdieu pointed out, this is "an interactionist vision... eliminating all structural effects and objective power relations" (Bourdieu 2005 [2000]: 198). The analysis in terms of "social networks" draws on the metaphor of the body to imply the notion of self-regulation of the social entity, and on "graphic reason" underpinned by the theory of graphs, to provide an intuitive interpretation of the formalization of relations. The "social network", "technical-economic network" and "public policy network" concepts, of use to the theorization of practices and to empirical studies, remain an epistemological subject of debate, for they are based on the metaphors of the technology of the reticular mind. That is why many authors have spoken of a "rainbow concept" (Bressand and Distler 1995: 184),[1] a "meso-concept" and even a "conceptual disaster" (Bott 1971 [1957]: 319]).

The degradation and dilapidation of the network concept in the technology of the mind creates not a void but an over-abundance of uses, discourses, notions, images and metaphors. A list of the uses of the notion in the various disciplines is enough to shed doubt on its coherence and consistency. It was short-lived in Saint-Simonian philosophy, where it served to conceptualize the shift of the network observed by physicians, in its natural state as an effect in and on the body, to the artificial network, calculated and constructed by engineers to cover a territory. The concept had hardly been formulated when the Saint-Simonians popularized it and turned it into a fetish, a cult of the technical network, surrounded with images and oriflammes. This excess of metaphorical uses seems to condemn it, as if the proliferation of its uses "in extension" entailed a void "in understanding", or even its dilution. We could conclude, on the contrary, that the use of a notion is proof of its effectiveness. Let us therefore draw up an inventory of the contemporary remains of the network concept. Its analysis enables us to distinguish several entangled meanings that we have arranged around two general forms: first, a techno-utopia, a set of narratives and fictions associated with technical networks; and second, a technology of the mind, which is a dispersion of the original concept. The fragments of the network concept remain a technical and technological pair: a technical matrix with its sack of metaphors, and a concept degraded into a technology of the mind. This duality is intrinsic to the contemporary concept, the functioning of which can be described as the passage between opposites, like a mediation in a contradiction, a path between places. The network can be defined only by relations, links or interconnections, considered as such, but it is always a networking of elements, links between places, an interconnection of entities. It links up that

---

[1] My own translation.

which was formerly isolated, separate or split up. This is the starting point of Michel Serres' definition of a network diagram constituted at a given point in time, "a plurality of points (summits) linked to one another through multiple ramifications (paths)" (Serres 1969: 11). The summit of a network is the intersection of several paths and, reciprocally, a path links up several summits. Michel Serres laid down the foundations of an epistemology of the network concept, by contrasting it with the linearity of the dialectical pathway. Henri Lefebvre likewise maintained that "while trellises and semi-trellises of multiple paths from each point to every other point (and in an infinite number of paths)... The trellis implies and allows for a different, more complex rationality" (Lefebvre 1969: XLVIII–XLIX). The network is richer than the tree – a particular type of network – and accounts for the complex, or the complex system, as shown by its history which has successively enabled us to interpret the invisible structures of the body, nature or society. This is the first characteristic of the concept: the network is non-linear; it defines a hidden order in a formalizable structure (diagram, graph, matrix) that enables us to explain the behavior of a complex system. Consubstantial to any conceptualization of the network is the hypothesis that there is a causal link between the structure and the behavior of a system, via the network. The representation of non-linearity and feedback in the network is especially complex because its topology, that is, its connectivity, changes with its "functioning". Here lies the heuristic advantage of "networks of automats" which are theoretical or formal models for explaining complex systems. While the network model enables us to make up for a shortcoming in the theory, it can, in turn, become an epistemological obstacle by imposing the network image on the disciplines that draw on it. Based on the example of networks of automats, Henri Atlan argues that, although "automata theory is used where self-organization translates into changes in the structure of a network, produced by the very functioning of the network" (Atlan 1992: XIV), this analysis of self-organized systems cannot be grasped fully through the prism of networks, for they yield only some of the knowledge of the hyper-complexity of the living. The value of a network is its possible modelling-formalization in graphic form, and the fact that it is more than a machine but less than the living, more than the linear but less than the hyper-complex, more than a tree but less than smoke (Atlan 1979). It functions as an intermediary being, its essence is "inter"; it is what connects. The substance of the network is the intermediary and the passage. The network marks the passage, without offering a limit or a precise boundary to its extension – unlike the system. The network appears at once as a degraded concept, a methodological precept, a graphic percept, and theoretical decept.

Despite this difficulty we have not given up on defining it, as a structure of unstable interconnections, composed of interacting elements, and whose variability is governed by a rule of functioning. There are three levels to the definition that I propose:

1. The network is a structure composed of interacting elements, the peaks or nodes, linked to one another by paths or links in a three-dimensional or "tabular" space. From this point of view crystal is a "pure network".

2. The network is a structure of interconnection that is unstable over time, for the birth of a network (the growth from element to network) and its transition from simple to complex, are consubstantial with its definition. The structure of the network includes its dynamics, as the modification of the network can be considered in a deterministic or random way. Whether we consider the development of one element into a whole network, or that of a network into a network of networks, we are always considering a complexity that is self-generated or even "self-organized" by the network's own internal dynamics. An organism would thus be a network in a natural state.
3. The modification of the structure is governed by a rule of functioning. It is assumed that the network structure's variability follows a norm, which it may be possible to model. This explains the functioning of the system structured as a network. We go from the network dynamics to the functioning or behavior of the complex system, as if the former were the invisible side of the latter, and therefore its explanatory factor.

These three aspects are confused in the contemporary notion of network, the essential virtue of which is to allow one or several elements to change into a whole; a whole defined at time t into the same whole at time t'; a hidden whole into a visible one. In other words, the network is at once the link between one element and an entity, between various states of an entity, and between the structure of an entity and the behavior of another entity. Thanks to the network, everything is a link, transition or passage, to the point of confusing the levels that it connects: whether interaction between elements, the engendering of one structure by another, or the functioning of a complex system.

Thus, two postulates are implicitly set with regard to the systematic use of the reticular model and which explain its success: the establishment of a causal link between the reticular structure of a complex system and its behavior; and the fragmentation and decomposition of the entity under study (nature, society, and organization) into atomized elements intended to be interconnected through the network. A determinism of the structure and a teleology of the system frame network thinking, between invisible order and visible complexity. The network provides an answer to the prior fragmentation of the object submitted to it; it links up that which has been separated.

The network is neither the system, nor the structure, and even less so the "rhizome" of Deleuze and Guattari (1980). The network is less constituted than the system and the structure, but it affords the advantage of encompassing the image of unlimited interconnection, a quasi-viral extension, as in the metaphor of the body and its natural regulation. The network is less organized than the system and remains open, but it allows us to conceptualize the behavior of the complex system and to find an explanatory formalization for it. The network always serves as a hidden structure of the complex visible system; one that can be modelled. It gives it support and form. As for the rhizome, it is not a network either, even if the two share some common features, especially the fact of hiding something. But while the network hides nothing other than its logical architecture, the *rhizome* hides an unconscious element inherent to the link. The *rhizome* remains unconscious and cannot be unfolded, unlike the network. But the strength of the notion of network lies less in its logic or graphic reasoning than in the cortège of images accompanying it. Metaphorically speaking, the network is always linked to its distant relationship with the organism, even though it has a source of regeneration in the development of every new technical communication network. The ambivalence of life (circulation of flows, the network is functioning) and death (breakdown, the network no longer functions) is consubstantial with this metaphor and the notion of network. The symbolism of the network is structured by the life/death pair, in which the network is sometimes identified with failure, suffocation or death, and sometimes with life. Depending on the mode of functioning of the network, we are on one side or the other; the network metaphor is two-headed: surveillance of the circulation and circulation of the surveillance. The network figure, like the Greek *métis*, is always ready to be inverted: from circulation to surveillance, or from surveillance to circulation. The network reactivates the image of the net that captures and lets escape, that holds back or releases... The network is within this basic surveillance-circulation pair and identified with one of the two terms, by opposition to the other. It also allows for moving from one term of the original opposition to the other, in the form of transition. The network can be either of the two terms of the symbolic grid-circulation opposition, and the rational medium term linking the two. Hence, the metaphorical work and the rational model of the network complete one another in the technology of the network mind, just as techniques and images combine in the techno-utopia. This set constitutes the *retiology* that is prevailing everywhere, including in the political sphere, on the Internet and in political parties – whether "Pirate" or not – that endeavor to organize themselves into networks or with networks. The question is now: is it possible to exit *retiology* and, if so, under what conditions? This critique is the prerequisite for any construction of another, post-cybernetic imaginary.

References

Atlan, H. 1979. *Entre le cristal et la fumée: Essai sur l'organisation du vivant, Coll. "Points-Sciences"*. Paris: Editions du Seuil.
———. 1992. *L'organisation biologique et la théorie de l'information*. Paris: Hermann.
Bott, E. 1971 [1957]. *Family and social networks,* 2nd ed. New York: The Free Press.
Bourdieu, P. 2005 [2000]. *The social structures of the economy*. Trans. C. Turner. Cambridge: Polity Press.
Bressand, A., and C. Distler. 1995. *La planète relationnelle*. Paris: Flammarion.
Deleuze, G., and F. Guattari. 1980. *Mille plateaux: Capitalisme et schizophrénie*. Paris: Éditions de Minuit.
Lefebvre, H. 1969. *Logique formelle, logique dialectique,* 2nd ed. Paris: Anthropos.
Serres, M. 1969. *Hermès I. La communication*. Paris: Editions de Minuit.

About the Authors

Steven Dorrestijn is a senior researcher in the Ethics and Technology group at Saxion University of Applied Sciences, the Netherlands. In 2012 Dorrestijn completed his PhD thesis (*The design of our own lives: Technical mediation and subjectivation after Michel Foucault*) at the University of Twente, the Netherlands. Previously he studied Philosophy in Paris and Philosophy and Mechanical Engineering in Twente. Dorrestijn's research and publications focus on the philosophy and ethics of technology, Michel Foucault's work in relation to technology, and the integration in design of knowledge about the impact of technology (Product Impact Tool). www.stevendorrestijn.nl

David Fernández-Quijada is Senior Media Analyst at the Media Intelligence Service of the European Broadcasting Union. His research focuses on public service media, media industry and markets, radio and media policy. He often delivers presentations on these topics at trade events and workshops. With a PhD in Communication and Media Studies, he was previously lecturer at the Autonomous University of Barcelona. His track-record includes more than fifty research publications. He is currently the vice-chair of the Media Industries & Cultural Production section of ECREA, the European Communication Research and Education Association.

José Luís Garcia is a Senior Research Fellow at the Instituto de Ciências Sociais, Universidade de Lisboa (ICS-ULisboa). He received his PhD in Social Sciences from the same University after pursuing doctoral studies at the London School of Economics and Political Science, University of London. His main research interests are the social theory of technology, sociology of technology, biotechnology, commodification and ethics, communication and information technologies. His most recent publications include a co-edition of *La Contribution dans l'Univers des Médias Numériques: Pratiques participatives à l'ère du capitalisme informationnel* (Presses de l'Université du Québec, 2014) and *Jacques Ellul and the Technological Society in 21st Century* (Springer, 2013). He is now finishing a book on *Technodicy*.

Pedro Xavier Mendonça received his PhD in Social Sciences-General Sociology from the Instituto de Ciências Sociais, Universidade de Lisboa (ICS-ULisboa). He was a PhD visiting student at Lancaster University. He is an assistant professor of the School of Business Communication, EFAP Portugal, and research collaborator at ICS-ULisboa. His main research interests are the relationship between information and communication technologies and rhetoric, the effects of enchantment in technology, material and traditional semiotics, innovation, and organizational communication. He has had several articles published in these fields.

Pierre Musso took a degree in Philosophy at the *École Nationale Supérieure des Postes et Télécommunications*, before obtaining a doctorate in Political Science at the University of Paris (1) Panthéon-Sorbonne, where he would later teach. Musso is professor of Information and Communication Sciences at the University of Rennes (2) and at *Télécom Paris Tech*, where he holds the education and study chair in "Modelling imaginaries, innovation and creativity" (*Modélisations des imaginaires, innovation et création*). From the body of his work it might be highlighted *Télécommunications et philosophie des réseaux: la postérité paradoxale de Saint-Simon* (PUF 1997), *Saint-Simon et le Saint-Simonisme* (PUF 1999), *Critique des réseaux* (PUF 2003) and the co-edition of the monumental critical edition of Henri Saint-Simon's *Oeuvres Complètes* (PUF 2012).

Francisco Rüdiger received a PhD in Social Sciences from São Paulo University. He teaches Critical Communication Studies and Philosophy of Technology at the Pontifical Catholic University and at the Federal University of Rio Grande do Sul, Porto Alegre, Brazil. Some books he authored are *O mito da agulha hipodérmica e a era da propaganda* [*The myth of hypodermic needle and the era of propaganda*] (2015), *O amor e a mídia* [*Love and the media*] (2013), *As teorias da cibercultura* [*The cyberculture theories*] (2010), *Martin Heidegger e a questão da técnica* [*Martin Heidegger and the question of technology*] (2008), and *Theodor Adorno e a crítica à indústria cultural* [*Theodor Adorno and the critique of culture industry*] (2003).

Rodrigo Saturnino received his PhD in Sociology from the Instituto de Ciências Sociais, Universidade de Lisboa (ICS-ULisboa). He was a PhD visiting student at University of São Paulo (USP), Brazil. He is a research assistant at the Centre for the Study of Migration and Intercultural Relations (CEMRI-UAB) and Media Analyst at Regulatory Authority for the Media (ERC). His main research interests are the formation of new identities and new practices based on digital media. His is now finishing a book based on his dissertation about "The Politics of Pirates".

Filipa Subtil is currently assistant professor at the School of Communication and Media Studies, Lisbon Polytechnic Institute, Portugal. She holds a PhD in Social Sciences with a thesis on US communication theory. Her research interests focus mainly on sociology of communication, the social theory of the media in the US and

Canada and the problematic frameworks of the media on gender issues, all fields on which she has published articles and essays. Subtil is author of *Compreender os Media: As Extensões de McLuhan* [*Understanding the Media: The McLuhan Extensions*] (MinervaCoimbra, 2006).

Dazhou Wang is a Professor at the School of Humanities and Social Sciences, University of Chinese Academy of Sciences. He received his PhD in Philosophy of Science and Technology from Northeastern University, China. His main research interests are the social and philosophical studies of technology and engineering, the Internet and society, innovation and research policy. His main publications include *Technology, Engineering and Philosophy* (Science Press, 2013) and *Knowledge, Fields and Innovations* (Chinese Social Science Press, 2005), both in Chinese. He is now writing a book on the history of the large-scale scientific engineering projects in China.

Kaixi Wang is a graduate student at the School of Humanities and Social Sciences, University of Chinese Academy of Sciences. She received her BA in Social Science from Beijing Language and Culture University in 2012. Her main research interests are social studies of the Internet and the development of e-philanthropy in China.

MIX
Papier aus verantwortungsvollen Quellen
Paper from responsible sources
FSC® C105338

If you have any concerns about our products,
you can contact us on
ProductSafety@springernature.com

In case Publisher is established outside the EU,
the EU authorized representative is:
**Springer Nature Customer Service Center GmbH
Europaplatz 3, 69115 Heidelberg, Germany**

Printed by Libri Plureos GmbH
in Hamburg, Germany